This is the propety of RM

EDEXCEL GCSE SCIENCE

Additional Science Student Book

Series Editor: Mark Levesley

Aaron Bridges, Ann Fullick, Richard Grime, Miles Hudson, Penny Johnson, Sue Kearsey, Jim Newall, Damian Riddle, Sue Robilliard, Nigel Saunders

ResultsPlus **authors:**

Mark Grinsell

Sue Jenkin

Ian Roberts

Julia Salter

David Swann

A PEARSON COMPANY

Contents

How to use this book

Progress questions can be used to check understanding as you work through the course.

Maths skills boxes appear throughout the Student Book to give you opportunities to refresh your maths skills.

Learning outcomes are taken straight from the specification to make it clear what you need to know for the exam.

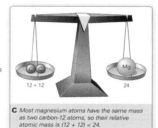

C2.3 The modern periodic table

How are the elements arranged in the modern periodic table?

Mendeleev's periodic table gradually became accepted by scientists. With a few adjustments as new discoveries were made, it led directly to the modern periodic table. This is now so familiar that you see it on t-shirts and mugs. You can even sit down for a meal at one (see Figure A).

1 What are the atomic number and mass number of an element?

For many years, an element's atomic number was just its position on the periodic table. Then, in 1913, scientists worked out that an element's **atomic number** was the number of protons in its atoms. The total number of protons and neutrons in the nucleus of an atom is called its **mass number**. In full chemical symbols like $^{27}_{13}Al$ the top number is the mass number and the bottom one is the atomic number.

B mass number (number of protons + number of neutrons)

$^{27}_{13}Al$

atomic number (number of protons)

This information can be used to calculate the number of protons, neutrons and electrons in an atom. For example, in $^{27}_{13}Al$, there are:
- 13 protons (because the atomic number is 13)
- 27 − 13 = 14 neutrons (mass number minus atomic number)
- 13 electrons (the same as the number of protons).

2 Work out the number of protons, neutrons and electrons in an atom of $^{23}_{11}Na$.

Relative atomic mass

The mass of an atom is incredibly small, so we use **relative atomic mass** instead of its actual mass in kilograms. The symbol for relative atomic mass is A_r. Carbon is used as the standard atom. A carbon-12 atom contains 6 protons, 6 neutrons and 6 electrons. Its relative atomic mass is defined as exactly 12. The masses of all other atoms are compared to this. For example, the mass of a helium-4 atom is one-third that of a carbon-12 atom. So its relative atomic mass is 4.

Maths skills

To find the relative atomic mass of chlorine you need to calculate the average (**mean**) mass of its atoms, where

$$mean = \frac{sum\ of\ values}{number\ of\ values}$$

C Most magnesium atoms have the same mass as two carbon-12 atoms, so their relative atomic mass is (12 + 12) = 24.

12 + 12   24

3 How are the elements arranged in the modern periodic table?

4 Describe what groups and periods in the periodic table are.

The modern periodic table

The elements in the modern periodic table are arranged in order of increasing atomic number rather than in order of atomic mass as Mendeleev did. The horizontal rows are called **periods**; the vertical columns are called **groups**. Each group contains elements with similar properties. The main groups are numbered 1 to 7 from left to right. The group on the far right is group 0.

122

D The modern periodic table shows th

Isotopes

Isotopes are different atoms and electrons but different nu always have 17 protons, but se neutrons. These are different

The presence of isotopes mea whole numbers. For example, 75% of all chlorine atoms, and the relative atomic mass of ch

relative atomic mass = (75

= 262

Learning outcomes

1.2 Classify elements as metals and
1.8 Explain the meaning of the term
1.9 Describe the arrangement of ele number, arranged in rows called
H 1.10 Demonstrate an understanding t
H 1.11 Calculate the relative atomic mas
HSW 14 Describe how scientists share da revise scientific theories

Higher-tier only questions and outcomes are clearly identified with small **H** icon.

These pages give you an investigatio task, allowing you to complete a specification practical. These tasks can be used as practice Controlled Assessments.

The specification practical is listed in the Learning outcomes box.

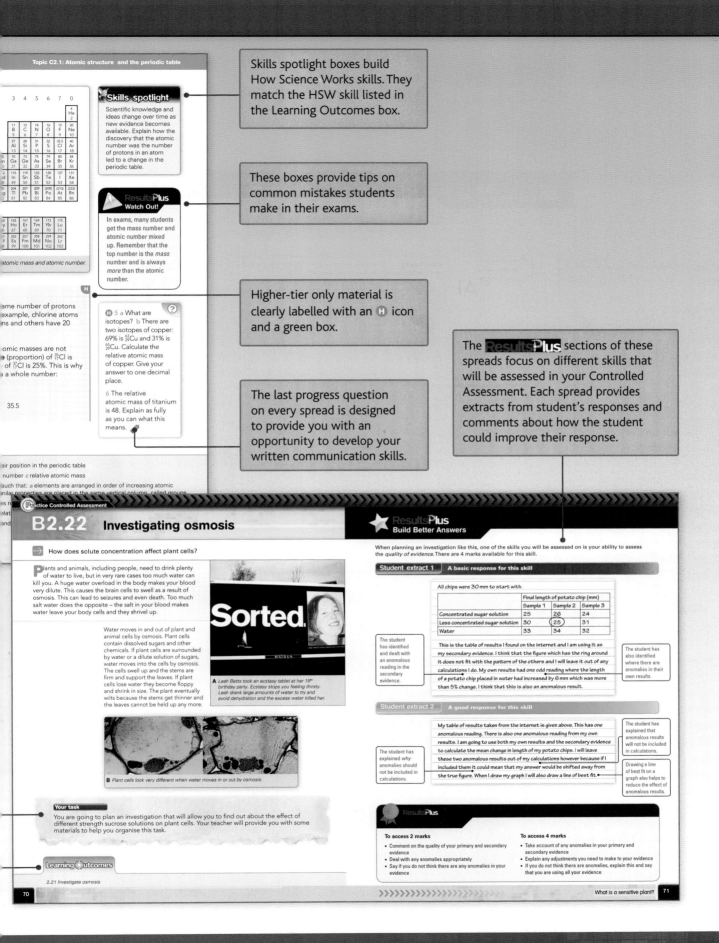

Skills spotlight

Scientific knowledge and ideas change over time as new evidence becomes available. Explain how the discovery that the atomic number was the number of protons in an atom led to a change in the periodic table.

Skills spotlight boxes build How Science Works skills. They match the HSW skill listed in the Learning Outcomes box.

ResultsPlus
Watch Out!

In exams, many students get the mass number and atomic number mixed up. Remember that the top number is the *mass* number and is always *more* than the atomic number.

These boxes provide tips on common mistakes students make in their exams.

Higher-tier only material is clearly labelled with an **H** icon and a green box.

H 5 a What are isotopes? b There are two isotopes of copper: 69% is $^{63}_{29}Cu$ and 31% is $^{65}_{29}Cu$. Calculate the relative atomic mass of copper. Give your answer to one decimal place.

6 The relative atomic mass of titanium is 48. Explain as fully as you can what this means.

The last progress question on every spread is designed to provide you with an opportunity to develop your written communication skills.

The **ResultsPlus** sections of these spreads focus on different skills that will be assessed in your Controlled Assessment. Each spread provides extracts from student's responses and comments about how the student could improve their response.

...ame number of protons ...example, chlorine atoms ...ns and others have 20

...omic masses are not ...e (proportion) of ^{35}Cl is ...of ^{37}Cl is 25%. This is why ...a whole number:

35.5

...ir position in the periodic table

...number c relative atomic mass

...such that: a elements are arranged in order of increasing atomic ...milar properties are placed in the same vertical column, called groups

---

Practice Controlled Assessment

B2.22 Investigating osmosis

How does solute concentration affect plant cells?

Plants and animals, including people, need to drink plenty of water to live, but in very rare cases too much water can kill you. A huge water overload in the body makes your blood very dilute. This causes the brain cells to swell as a result of osmosis. This can lead to seizures and even death. Too much salt water does the opposite – the salt in your blood makes water leave your body cells and they shrivel up.

Water moves in and out of plant and animal cells by osmosis. Plant cells contain dissolved sugars and other chemicals. If plant cells are surrounded by water or a dilute solution of sugars, water moves into the cells by osmosis. The cells swell up and the stems are firm and support the leaves. If plant cells lose water they become floppy and shrink in size. The plant eventually wilts because the stems get thinner and the leaves cannot be held up any more.

A Leah Betts took an ecstasy tablet at her 18th birthday party. Ecstasy stops you feeling thirsty. Leah drank large amounts of water to try and avoid dehydration and the excess water killed her.

B Plant cells look very different when water moves in or out by osmosis.

Your task

You are going to plan an investigation that will allow you to find out about the effect of different strength sucrose solutions on plant cells. Your teacher will provide you with some materials to help you organise this task.

Learning Outcomes

2.21 Investigate osmosis

ResultsPlus
Build Better Answers

When planning an investigation like this, one of the skills you will be assessed on is your ability to assess the *quality of evidence*. There are 4 marks available for this skill.

Student extract 1 | **A basic response for this skill**

All chips were 30 mm to start with.

	Final length of potato chip (mm)		
	Sample 1	Sample 2	Sample 3
Concentrated sugar solution	25	26	24
Less concentrated sugar solution	30	25	31
Water	33	34	32

The student has identified and dealt with an anomalous reading in the secondary evidence.

This is the table of results I found on the internet and I am using it as my secondary evidence. I think that the figure which has the ring around it does not fit with the pattern of the others and I will leave it out of any calculations I do. My own results had one odd reading where the length of a potato chip placed in water had increased by 6 mm which was more than 5% change. I think that this is also an anomalous result.

The student has also identified where there are anomalies in their own results.

Student extract 2 | **A good response for this skill**

My table of results taken from the internet is given above. This has one anomalous reading. There is also one anomalous reading from my own results. I am going to use both my own results and the secondary evidence to calculate the mean change in length of my potato chips. I will leave these two anomalous results out of my calculations however because if I included them it could mean that my answer would be shifted away from the true figure. When I draw my graph I will also draw a line of best fit.

The student has explained *why* anomalies should not be included in calculations.

The student has explained that anomalous results will not be included in calculations.

Drawing a line of best fit on a graph also helps to reduce the effect of anomalous results.

ResultsPlus

To access 2 marks
- Comment on the quality of your primary and secondary evidence
- Deal with any anomalies appropriately
- Say if you do not think there are any anomalies in your evidence

To access 4 marks
- Take account of any anomalies in your primary and secondary evidence
- Explain any adjustments you need to make to your evidence
- If you do not think there are anomalies, explain this and say that you are using all your evidence

What is a sensitive plant?

There is a practice exam paper for both the Foundation and Higher tiers in each unit.

Each exam section has 2 extended writing questions where you can practise answering this type of question. The extended writing questions are worth 6 marks and are always the last part of questions 4 and 5.

The question parts are colour coded to indicate what grades they can access:
- Orange means you can access grades G-D (Foundation tier) or D-B (Higher tier)
- Light green means you can access grades E-C or B-A*
- Dark green covers the whole grade range, G-C or D-A*.

The Build Better Answers pages present an extended writing question along with three different student answers to the question – a level 1, a level 2 and a level 3 answer.

ResultsPlus Exam Practice

Foundation tier

The digestive system

5. The diagram shows the human digestive system.

mouth
oesophagus
liver
stomach
pancreas
small intestine
large intestine

(a) (i) Draw **one** line from each part of the digestive system to its correct function. One has already been done for you.

Part of digestive system | Function

mouth

large intestine

liver

produces bile

produces three types of digestive enzymes

contains saliva to moisten food

forms faeces from undigested food

(2)

(ii) Describe the role of the stomach as part of the digestive system. (2)

(iii) Describe how the oesophagus helps to move food from the mouth to the stomach. (2)

(b) Richard eats a cheese sandwich for his lunch. Explain what happens to the sandwich as it moves through the various structures in Richard's digestive system. Use the diagram of the digestive system to help with your answer. (6)

100

Cloning the woolly mammoth

1. Scientists have unearthed body cells of a woolly mammoth that had [been frozen] for thousands of years. These cells contain frozen DNA samples that [could be used to clone] the woolly mammoth.

(a) (i) The woolly mammoth has been extinct for over 13 000 years. [One problem with] cloning a woolly mammoth is that

A there are no live sperm cells to fertilise the female egg
B the frozen DNA contains the haploid number of chromosomes
C the body cells found contain the diploid number of chromosomes
D there is no suitable surrogate to implant a re-nucleated egg

(ii) The diagram shows how a parent cell containing the DNA of a [woolly mammoth] divides to produce two new cells.

parent

(iii) Describe how [...]produce the t[...]
(iii) Describe how [...] a woolly mam[...]
(iv) Describe the s[...]

ResultsPlus Build Better Answers

Here are three student answers t[...] comments around and after them[...]

Question | **Stru[...]**

The diagrams show the structu[...] carbon, diamond and graphite[...]

Use the diagrams of the structu[...] graphite to help you explain w[...] electricity but diamond does ne[...]

Student answer 1 | **Extr[...]**

Examiners expect answers to be written in good English. If your English is poor overall you may lose marks.

Diamonds ge[...] of strong bo[...] Diamond don[...]

Examiner sum[...]
The answer has de[...] there are delocali[...] not relate these e[...]

Saying that dia[...] diagram, and it [...]

Student answer 2 | **Extr[...]**

Good: the electrons are referred to as 'free', and the answer goes on to say that they are able to move. However, 'delocalised' would be even better than 'free'.

Graphite is m[...] electrons. Wh[...] electrons ca[...] diamond beca[...] strong bonds[...] strong bonds[...]

Examiner sum[...]
This answer shows[...] electric current an[...] why diamond does[...]

200

Exam practice spreads provide a large bank of the new question types you will encounter in your exams.

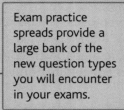

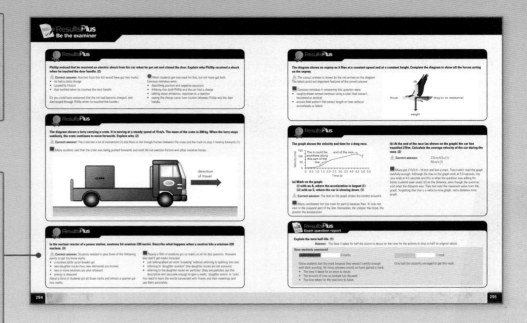

The Be the Examiner spread offers a variety of past exam questions along with comments, advice on how to improve an answer and common mistakes that students have made in the past.

(1)

There are comments around and under each student response, making it clear how to progress through the mark scheme to improve your response.

There is also advice on the things you can improve in your answer to help you move up a level in your response.

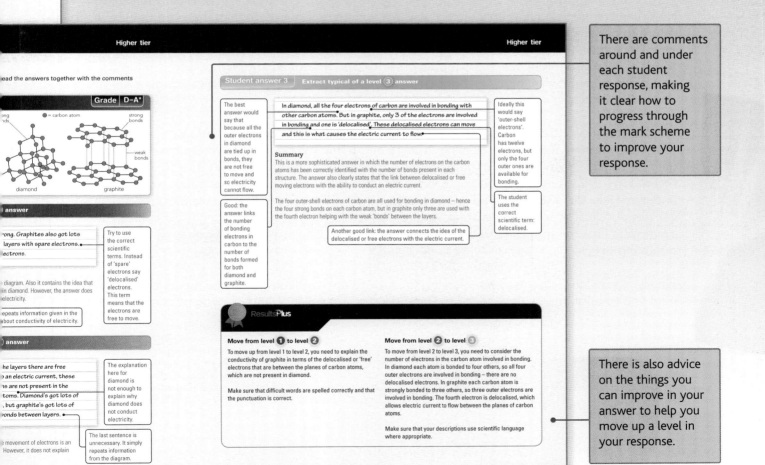

Why study science?

If you asked your science teacher why it's important to study and understand science, they may say, 'Everything around you is science.' You might expect a scientist to say that, but there's truth in it.

Just think of what your daily life would be like without advances in science. Even over the last 100 years, the achievements of science have benefited the human race in so many areas: in transport (cars, planes), in medicine (anaesthetics, organ transplants, cures for many diseases, drugs), in electronics (computers, mobile phones), in materials (fabrics such as Gore-Tex ® and Kevlar ®) and in agriculture (fertilisers, pesticides, genetic modification) to name but a few.

However, sceptics will say these advances have come at a price. Alfred Nobel, whose will funds the Nobel Prizes, invented dynamite for the mining industry, yet it rapidly became part of the armaments industry. And a hundred years ago, no one had heard of global warming or the ethical arguments over gene therapy. But in order to understand whether these sceptics are right, it's important that you learn some of the science behind these issues.

So why study science? Firstly, there is the excitement of discovery. Although you may not discover anything new yourself, you'll be able to experience some of the 'buzz' of the great scientists by working out some key ideas for yourself.

Science also gives you the chance to do some practical work: to get your hands dirty with experiments. You can investigate living organisms, create chemical reactions that give off energy or make colour changes, or build a physical model of a phenomenon.

However, learning about science in the new millennium is about more than that – and this is where we return to our sceptics. You need to be able to make informed decisions about how science benefits your life and what to do in situations where science has posed as many questions as it has answered. Much of what you might see in the media presents one side of an argument – and is often designed as a piece of journalism, high on shock value and low on facts! Knowing some facts yourself will enable you to look critically at how science is presented in the media.

It's also important to realise the importance of science to your future. Scientists work in so many fields to try and improve the life and health of the planet. Some will be in research, perhaps working on the cure for cancer or HIV / AIDS, or helping make the chips to power the next generation computer-game technology. Others may be working in industries trying to develop new energy sources or new fabrics and materials. But you'll also find science specialists working as weather forecasters, television researchers, lawyers, medical specialists, teachers, writers, architects, journalists and in many other areas. Increasingly, employers look for applicants with a science background because they know they will have logical, enquiring minds.

Finally, there's one other excellent reason to study science: it's fun! We hope that this book will help you throughout GCSE Science to see the interest, relevance and enjoyment of the subject.

Good luck!

Damian Riddle
Science Team
Edexcel

The units

At this stage, you'll either have covered the content for Unit 1 and Unit 2 in each of Biology, Chemistry and Physics in anticipation of taking separate science GCSEs at the end of your course, or you'll have achieved Science and Additional Science GCSE already and are now moving on to complete Further Additional Science.

Each of the periods of study of Unit 3 will lead to a written examination in one of the sciences: one each in Biology (B3), Chemistry (C3) and Physics (P3). In addition, you will have to complete Controlled Assessment. If you're studying for Biology, Chemistry and Physics, you will need to take a Controlled Assessment for each science. It is possible for you to take the Controlled Assessment tasks from either Unit 2 or Unit 3, so you may have already done this section. If you're studying for Further Additional Science, you will only take one Controlled Assessment, similar to the one you completed for Additional Science, but taken from Unit 3. Simply, you'll have to develop an idea or hypothesis to test, collect some data, process it and draw conclusions.

To remind you, here's the diagram to show you how the GCSEs available in the science subjects all fit together.

You're probably used to the fact that each of the Units is broken down into smaller topics. Let's start, therefore, by looking at the smaller topics that make up the Units you'll study as you complete the GCSEs in Biology, Chemistry and Physics, or Further Additional Science.

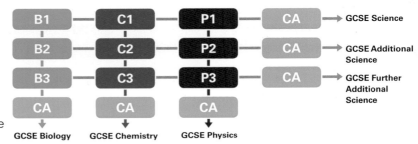

Unit B2

This unit, called 'The components of life', looks in detail at the way in which plants and animals have developed structures and functions to keep them alive. It is split into three smaller topics.

In Topic 1, you will:
- investigate the structure of cells in animals, plants and bacteria, and DNA
- see how scientists use genetic engineering and cloning to develop organisms with desirable characteristics
- discuss the ways in which cells divide during growth and reproduction
- explore how enzymes work.

In Topic 2, you will:
- learn how plants and animals get their energy through respiration
- explore the processes of photosynthesis and osmosis in plants
- use fieldwork to gain an understanding of ecosystems.

In Topic 3, you will:
- consider evidence for evolution
- explore the two main systems that keep humans alive: digestion and blood circulation.

Unit C2

The title of this unit – 'Discovering Chemistry' – reflects its focus on key areas that make up the science of chemistry. There are six topics in total and these help you to study:

- the structure of atoms and how they are arranged in the modern Periodic Table
- how metallic and non-metallic elements can react to form ionic compounds, and how these ions can be identified through chemical tests
- how covalent compounds can form when non-metals react together and how some mixtures of substances can be separated
- three very different sets of elements in the periodic table: the reactive metals of group 1, the reactive non-metals of group 7 and the inert gases of group 0
- the energy changes in chemical reactions and factors that can speed up or slow down these reactions
- how chemists can measure the amount of products made in reactions.

Unit P2

Like the chemistry unit, the physics unit, 'Physics for your future', is also divided into six topics. These topics aim to show how physics concepts are used in practical ways in the world today. The six topics cover:

- how electrostatic charges develop and how a knowledge of electrostatics is useful when refueling aeroplanes or using a spray paint can
- the components of electrical circuits; how current, voltage and resistance are related in these circuits; and how electric circuits can transfer power and energy
- the idea that forces make objects move and accelerate and how this can be represented by graphs and how this relates to objects in free-fall
- the ways in which momentum is encountered by drivers and car designers; how some types of energy can be calculated and how energy relates to work and power the forms of radiation emitted by unstable atoms and how we harness the energy of atoms through nuclear fission
- some uses of radiation, especially in medicine, and how people use radioactive materials safely.

In the next few pages, we'll take an in-depth look at the exams you'll take in GCSE Science.

This will help you get familiar with what the papers look like and will also give you some information on the sorts of questions that you'll be asked, plus the way in which the examiners use key words to test your knowledge and understanding.

Assessment Objectives

Many students think that examiners have the job of trying to catch them out or that examiners set out to write exam papers which are too hard. Nothing could be further from the truth! Examiners try hard to write questions that allow you to show everything you have learned during your GCSE course. However, they also have to make sure that any exam paper is fair for you, past students and future students. The papers need to be the same difficulty and test the same skills.

To help with this, all papers are targeted to test specific skills called Assessment Objectives. They sound a bit complicated, but here's what they mean:

Assessment Objective	The jargon	What it means
AO1	Recall, select and communicate your knowledge and understanding of science.	Essentially, AO1 questions will ask you to write down facts that you have learned.
AO2	Apply skills, knowledge and understanding of science in practical and other contexts.	AO2 questions may ask you to apply what you've learned to new situations, or to practical contexts – it's about showing skills you've learned, rather than repeating facts.
AO3	Analyse and evaluate evidence, make reasoned judgements and draw conclusions based on evidence.	AO3 questions will be about how well you can use data, say what graphs show, or think about arguments with two sides, such as ethical issues.

Your exam papers will be written so that the balance of these AOs is roughly 45% AO1, 35% AO2 and 20% AO3. The Controlled Assessment task will be about 50% AO2 and 50% AO3.

Types of question

Your exam paper will always consist of six questions. The questions will examine different areas of the course which you have studied. Each question will start with some straight-forward question parts, which will slightly increase in difficulty until the end of the question. The six questions themselves will also be 'ramped' slightly in difficulty: this means that Question 6 will be a little more challenging than Question 1. However, each question will start with a question part that puts you at ease and settles you into the question as a whole.

The first two questions will be worth around 8 marks, the next pair around 10 marks and the final pair around 12 marks. Each question could contain the following types of question:

- **Multiple choice**. These ask a question and usually provide four possible answers: A, B, C and D. Some multiple choice questions may ask you to choose words from a box to complete a sentence or to draw lines between statements.

- **Open response**. Like multiple choice, these will usually be worth 1 mark, but you will need to write down your own answer, rather than choose from a selection of answers provided.

- **Short answers**. These questions, worth 2 or 3 marks, will ask you to write slightly longer answers (perhaps three or four lines). Occasionally, at Higher tier, some questions – most likely those involving calculations – may be worth 4 marks.

Two questions in the paper, usually questions 5 and 6, will also have a piece of extended writing.

- **Extended writing**. These questions are worth 6 marks. They will require you to write at greater depth and the questions will be slightly more open to give you the chance to express yourself. In these questions, you will also be assessed on the quality of your written communication.

There is more detail on each type of question, and how to answer them, on pages 14 and 15.

The language of the exams

An exam paper contains precise words. It's important that you understand exactly what the words are asking you to do so that you can answer the question quickly and accurately.

The words are called 'command words' and it's useful for you to know what they mean. Some are simple words designed to allow you to give simple answers, whereas others are more complex terminology, asking you to write at greater length and give more detail.

Let's have a look at the command words you might encounter on a GCSE Science paper. These are in order of complexity below, i.e. they get more difficult.

Give, Name, State	These questions are asking for short answers – often only a few words – but with precise use of scientific terminology.
Complete, Select, Choose	In these questions you are usually asked to fill the gaps in sentences, using words that are given to you.
Draw, Plot	These questions will ask you to put data onto a graph, or draw a diagram. Make sure that your work is accurate and that you use labels, if appropriate.
Calculate, Work out	In GCSE Science, you'll need to use your mathematical skills in solving equations and performing simple calculations. This type of question will test you on these skills.
Describe, Use the graph	In these questions, try to use scientific terminology concisely and accurately. You may be asked to describe a particular process – in which case, remember to think about putting your answers in a logical sequence. Alternatively, you may be asked to describe the trend in a graph – here you will get more marks for using data that you have read from the graph.
Suggest	These questions are often asking you to apply what you have learnt to a new situation, or to put together different facts you have learnt in order to answer the question.
Explain	These questions will be worth 2 or 3 marks. You need to give reasons for your answer and not to simply repeat information given in the question. Imagine that any question that starts 'Explain...' is really asking you 'Why...?' – that should help you understand the depth needed in your answer.
Evaluate, Discuss, Compare	These command words will often be found in the 6-mark extended writing questions. Look carefully at exactly what the question is asking you to do. In many cases, it will be asking you to look at 'pros' and 'cons' surrounding a particular argument, or to give a balanced answer, looking at different aspects of a topic. 'Compare' questions will ask you to look at areas of similarity and difference between two ideas, or two sets of data.

Knowing what to expect: the front page

We know it can feel a bit scary sitting in the exam room with your GCSE paper in front of you, waiting to start. If you know what the paper looks like beforehand and what some of the text on it means, hopefully it won't seem so daunting.

Usually, the centre number will be written clearly somewhere in the exam room. It's a 5-digit number, and it will also appear on the statement of entry that your teacher or form tutor may have given you. If you're not sure, just ask one of the invigilators.

This is a 4-digit number, unique to you in your school or college. It's important to copy this down correctly because the exam board will use this number to give you your marks.

All GCSE Science papers last one hour and contain 6 questions.

GCSE Science papers come in two tiers: Foundation (which allows you to score grades G – C) and Higher (which allows you to score grades D – A*). Check you've got the correct tier of paper. Your teacher should have talked to you about the tier you're sitting.

All GCSE Science papers are marked out of 60. The mark for each question part is immediately after that part and the total for the whole question is at the end of the question. Questions marked with an asterisk indicate that the examiners will look at how well you express yourself and use technical vocabulary in your answer.

The paper asks you to use a black ink pen or biro. Your exam paper is scanned and marked by a marker on a computer screen and black ink means the marker can read your answers more clearly. It's also important to answer in the spaces provided. Don't feel you always have to fill the entire answer space though!

There's also a section here giving you advice how to answer questions.

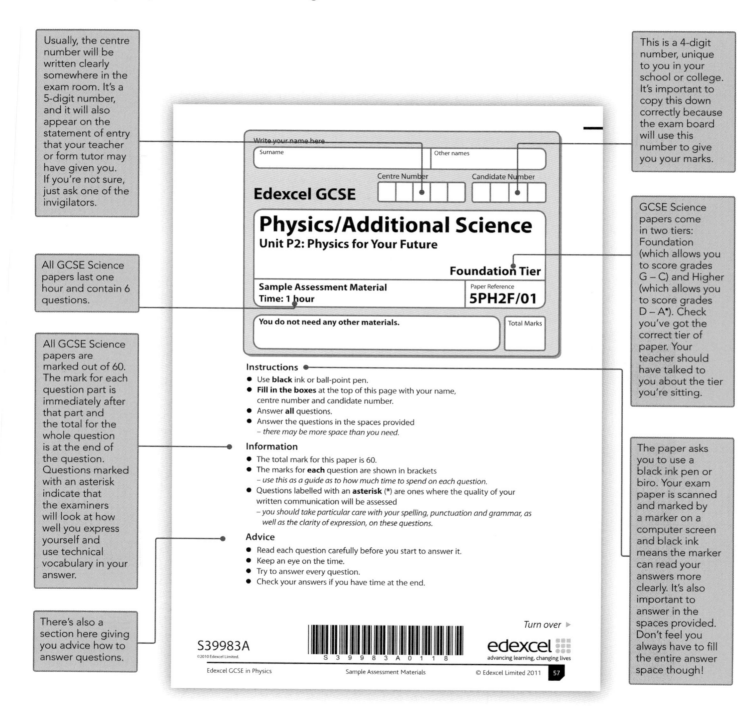

Write your name here

Surname

Other names

Centre Number

Candidate Number

Edexcel GCSE

Physics/Additional Science
Unit P2: Physics for Your Future

Foundation Tier

Sample Assessment Material
Time: 1 hour

Paper Reference
5PH2F/01

You do not need any other materials.

Total Marks

Instructions

- Use **black** ink or ball-point pen.
- **Fill in the boxes** at the top of this page with your name, centre number and candidate number.
- Answer **all** questions.
- Answer the questions in the spaces provided
  – *there may be more space than you need.*

Information

- The total mark for this paper is 60.
- The marks for **each** question are shown in brackets
  – *use this as a guide as to how much time to spend on each question.*
- Questions labelled with an **asterisk** (*) are ones where the quality of your written communication will be assessed
  – *you should take particular care with your spelling, punctuation and grammar, as well as the clarity of expression, on these questions.*

Advice

- Read each question carefully before you start to answer it.
- Keep an eye on the time.
- Try to answer every question.
- Check your answers if you have time at the end.

Turn over ▶

S39983A
©2010 Edexcel Limited.

S 3 9 9 8 3 A 0 1 1 8

edexcel
advancing learning, changing lives

Edexcel GCSE in Physics   Sample Assessment Materials   © Edexcel Limited 2011   57

Knowing what to expect: a sample question

This is called the stem of the question. It lets you know what topic the question is on. In some cases, the stem gives some important information to set the scene for the question.

Each question will contain what examiners call a stimulus. Most often, this will be a photograph or a diagram that should help set the context for the question. In some cases, the stimulus might be some data from an experiment or a short piece of text.

Here's an example of one of the command words that we looked at on page 11. It tells you what to do in your answer.

Examiners use bold text to highlight something they want you to notice. In many cases, it will be a word which you will be asked to define or it will tell you the number of answers you are expected to give.

The number of marks for the question part is given on the right-hand side of the exam paper, just under the question part. It's always a good idea to look at how many marks the question part is worth. This will give you some idea of how much you are expected to write, what sort of depth the examiners are looking for and the length of time you should spend on the question part.

Multiple choice questions are laid out differently – we'll look at these in more detail on the next page.

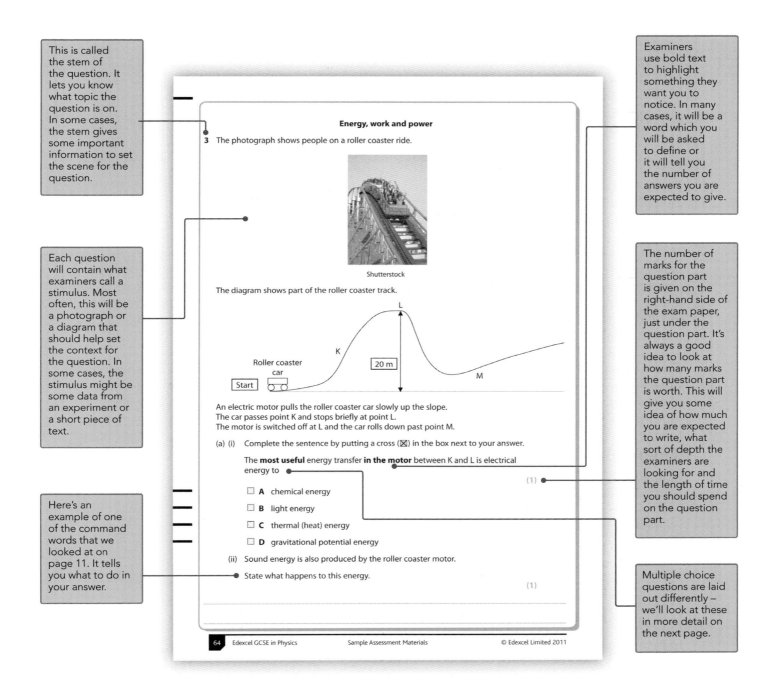

Energy, work and power

3. The photograph shows people on a roller coaster ride.

Shutterstock

The diagram shows part of the roller coaster track.

Roller coaster car

Start

20 m

K

L

M

An electric motor pulls the roller coaster car slowly up the slope.
The car passes point K and stops briefly at point L.
The motor is switched off at L and the car rolls down past point M.

(a) (i) Complete the sentence by putting a cross (☒) in the box next to your answer.

   The **most useful** energy transfer **in the motor** between K and L is electrical energy to

   (1)

   ☐ A chemical energy

   ☐ B light energy

   ☐ C thermal (heat) energy

   ☐ D gravitational potential energy

(ii) Sound energy is also produced by the roller coaster motor.

   State what happens to this energy.

   (1)

On pages 10 and 11, we looked at the types of questions that would come up in your GCSE Science exams. Let's look at each one in a bit more detail, so you understand how to score as many marks as you can.

Multiple choice questions

The most common type of multiple choice question will present you with four alternative answers, one of which is correct. For example:

> Which of these is an addictive substance contained in tobacco?
> A a cannabinoid
> B carbon monoxide
> C nicotine
> D tar

The best way to answer multiple choice questions is to cover up the four possible responses when you read the question. Think of your answer and then look at the four possible answers: hopefully yours will be among them!

If your answer isn't there – or if you couldn't come up with an answer when you looked at the question to start with – then you have to start looking at the answers one by one and eliminating ones that you know are wrong. Hopefully, you'll be able to eliminate three answers and be left with the correct one. It may be that you're only able to eliminate one or two wrong answers – but at least you can make a more educated guess. The key thing with a multiple choice question is this: don't leave it blank. You can't lose marks if you get it wrong and you may actually guess right!

There are similar questions where you're given the answers as part of the question. These might be selecting words from a box to complete sentences, or drawing lines to join words in one column to those in another. The same principle applies to these types as to the standard 'ABCD' multiple choice questions – look at the question before looking at the answers to see if you already know the answer.

Open response and short-answer questions

These questions might be 1 mark questions, often asking you to recall information. Or they may be worth 2 or 3 marks, if you're being asked for two reasons, or two factors.

Some examples of 1 mark questions might include:

> a Give one form of naturally-occurring calcium carbonate.
> b Which waves in the electromagnetic spectrum have the longest wavelength?

Other types of 1 mark questions might ask you to label or complete diagrams. Some may be of the **Suggest** type, asking you to use information that you already know and combine it with information in the question to come up with an answer.

Remember that these questions are only worth 1 mark – so don't spend too much time going into a great amount of detail. Keep your answer clear and concise. You don't have to answer in full sentences, so you can save time here.

Longer-answer questions

Each question on the paper will have some questions worth 2 or 3 marks. Generally, there will be more of these longer-answer questions towards the end of the exam paper. These types of question will include the following:

Writing or completing equations. Make sure with these that you're clear whether the answer needs a word equation or a chemical equation. Don't forget to look for clues in the question – you'll often find that the words or formulae you need are in the question.

Drawing graphs. Different questions will ask you to do different things with a graph. Read the question carefully to check if you're meant to label axes or draw your own axes. Be careful when you plot the points – there are marks for the accuracy of your plotting. Finally, check to see if you're required to draw a line of best fit – and remember, lines of best fit don't have to be straight lines, they can be curves!

Performing calculations. Two key tips for calculations. Firstly, always show your working. If you make a slip with your calculator (and end up with the wrong final answer), you can often get marks for your working if the examiner can see that you just made a small slip. Secondly, check to see if you need to give units in your answer.

Describe or **Explain** questions. For example, a question like:

> Describe the path taken by a nerve impulse from the receptors to the effectors.

These questions are designed to be a bit more difficult than some of the other types we've seen. However, there's no reason why you shouldn't score well on them. It's always best to spend a few moments planning your answer before starting to write. Make sure you understand exactly what you're being asked to do – you want to keep your answer relevant. Do think about the number of marks that the question is worth: if it's a 3 mark question, then you need to make three different points.

Extended writing questions

There is one final type of question that you'll face in your examination paper – the extended writing question. That may sound a bit daunting, but these questions aren't designed to be really difficult or to set traps for you. Instead, they give you more time and space to show the knowledge and skills you have picked up during your GCSE course. These questions will be written to test you on a variety of different skills.

The other important thing to realise about these questions is that they're not just aimed at A* students. The questions are designed to be open enough to allow all students to be able to write something.

On the next two pages we'll look at the sort of topics that might form the basis of an extended writing question.

Practical-based questions

Much of the time you spend in science lessons, you're doing practical work. Some extended writing questions will therefore ask you about the practical work that you've done. The questions may ask you about how you would set up an investigation or about how a particular practical investigation gives us information about scientific theories.

Opinion-based questions

Many people think that science is all about hard facts and certainties but many areas of science still lead to uncertainty. In some cases, this is because scientists genuinely don't know all the answers or because data that they collect can be interpreted in different ways. In other cases, the discoveries that scientists have made have led to moral or ethical dilemmas.

Opinion-based questions will therefore ask you to use what you have learnt to back up your own opinion or to show that you understand two sides of a particular argument.

Knowledge-based questions

There will be some parts of your GCSE Science course where you will learn quite a lot of detail. Sometimes, 1 or 2 mark questions on these topics mean you can only write about some of the things you know about the topic, rather than all of it. Equally, there are times when the examiners want to ask a general question to allow you to show a variety of things you know or to bring in knowledge across the whole subject.

Hence, some extended writing questions will ask you for more factual responses. This might seem a bit daunting, but the questions may sometimes be phrased in such a way as to allow a variety of answers. There will often be some prompt questions to help you get started.

How they're marked

The extended writing questions are always worth 6 marks. So if you've read the advice given for other questions types, you're probably thinking about the amount that you may have to write, and you're thinking that you have to make sure you make 6 credit-worthy points.

Well, you're half right – the whole point about these questions being called 'extended writing' is that there is the opportunity for you to write at greater length. Note that in many questions, you'll be able to provide diagrams or tables as part of your answer, so don't think it's all about writing essays! But you don't have to think about making six points each time. These questions will be marked with two ideas in mind – one is known as 'levels' and the other as 'QWC'.

Let's look at **levels** first. The mark scheme for each question contains a long list of possible points – the key thing to remember is that you're not expected to write all of these! The examiners will see lots of examples of answers, and they'll use this experience to place your answer in one of three levels: very good (Level 3), good (Level 2) and less good (Level 1). Their judgement will be based on factors such as the number of points you've made, the balance in any argument you've made or how well you've appreciated your practical work. Each level corresponds to 2 marks. Level 1 is 1–2 marks, Level 2 is 3–4 marks, and Level 3 is 5–6 marks. So once your answer is placed in one of the levels, how is the final mark decided?

The other deciding factor is **QWC**. This stands for Quality of Written Communication. What the examiners will look for is your ability to use proper scientific words correctly and use proper spelling, punctuation and grammar, your ability to put your answer in a logical order and how well you express yourself.

How to answer

Let's have a look at a possible question:

> Discuss the advantages and disadvantages of using biofuels instead of petrol for cars.

The question relates directly to a particular part of the specification – the statement about biofuels being renewable but taking up land that could be used for growing crops. However, your answer wouldn't have to involve that. You could, for example, talk about what factors make a good fuel and use this to compare biofuel and petrol; or you could write about the amount of carbon dioxide given off by each fuel and hence talk about the greenhouse effect. Or if you'd learned a lot about the effects of acid rain, you could even write about sulfur impurities in petrol causing this, whereas biofuels (which don't contain sulfur) don't give the same problems.

The key, therefore, is to treat each question as a way for you to display some of the knowledge you've learned, rather than just a black-and-white question with one set answer.

In terms of your exam technique, the most important things to do are: a) think about all the different areas of what you've learned that you can use in your answer, and b) to spend some time planning your answer. The two extended questions are worth a total of 12 marks between them – a fifth of the marks on the paper – so you should spend a reasonable amount of time on them, probably about 6–8 minutes on each. And remember that this is one area where you need to be careful with the logical presentation of your answer. Finally, read the question again before answering – did you notice that it asked for advantages and disadvantages? Remember, therefore, to think about both pros and cons in your answer.

Let's look at a possible answer to this question:

> Biofuel is a renewable fuel. It is often made from fermenting the sugar in sugar beet to produce ethanol. Petrol, however, is not renewable – it is a fossil fuel and, at the rate humans are using it, it will soon be used up. Petrol, however, does give more energy out when it burns than biofuels do. So, although biofuels may be an answer to the problems of energy needs, we'll need to burn a lot of them to get our energy. That means that a lot of land needs to be used to grow the sugar crop – and that means less land for growing food for people or livestock. Many people see biofuels as being "green". It is certainly true that, unlike petrol, they don't contain sulfur impurities, so there is no sulfur dioxide produced when they burn, so no acid rain. Biofuels do, of course, produce carbon dioxide when they're burnt, just like petrol, and this is a powerful greenhouse gas. But, some of the effect of this may be offset as, to grow the next crop of sugar beet, carbon dioxide will be taken in by the plants photosynthesising.

This answer looks at some pros and cons about using petrol. Importantly, it also looks at some negatives of using biofuels – such as land use – as well as seeing the positives. This means it should be a top-level response. The presentation of the answer is good, with accurate spelling and grammar, and the answer is logically structured. It should achieve a mark in the top level.

There is an aspect of your course where we try to assess you on your ability to handle practical work and use the results of experiments. This section is known as Controlled Assessment.

If you've got older brothers and sisters or friends in years above you at school, it's quite likely that they'll have told you horror stories about coursework and how much time it took up when they were studying for their GCSEs.

Well, Controlled Assessment is, essentially, a form of coursework, but one which is very different from the coursework used in the past.
Hopefully, you'll see that many of the differences are positive ones.

The major differences are:

- In Controlled Assessment, the task is set by the exam board and changes every year. This means that it is the same task for all students, which is much fairer.

- In Controlled Assessment, parts of the work have to be completed in class time and you won't be able to take work home. This means that it is easier to guarantee that it is your own work and means that students who have family members who are good scientists aren't able to get their family to help with their work.

- Controlled Assessment takes up less time – especially because you do not have to do write-ups for several homeworks.

How important is Controlled Assessment?

Controlled Assessment is worth 25% of your GCSE – in other words, it has the same proportion of marks as a written exam.

How many Controlled Assessment tasks do you need to do?

You have to submit one Controlled Assessment mark for GCSE Science, another one if you go on to GCSE Additional Science, and then another one if you go on to Further Additional Science. This can come from any of the three sciences: Biology, Chemistry or Physics, although if you do go on to Further Additional Science, your Controlled Assessment mark has to come from a different science to the one you submitted for your Additional Science Controlled Assessment. It is possible that your teachers may complete more than one Controlled Assessment task with you – if this is the case, you submit your best mark. In some cases, you may be able to combine marks for different Controlled Assessments in order to get a higher mark. Your teacher will be able to give you some advice on this.

If you are taking Biology, Chemistry and Physics as separate GCSEs, then you will need one Controlled Assessment for each science.

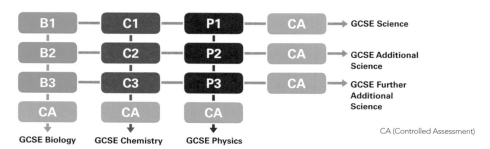

What sort of practicals will the exam board choose?

The specification for GCSE Science includes a series of suggested practicals. The Controlled Assessment could be set on one of these pieces of practical work, or it may be set on a related experiment. This is another example of us trying to ensure that the Controlled Assessment is as fair as possible. Having asked you to study some pieces of practical work in your course, it seems fair that one of these pieces of work – or something closely related to those pieces of work – will be used as the Controlled Assessment, as all candidates will have the same previous experience of the practical being used.

How will you prepare?

Preparation for the Controlled Assessment is obviously going to be very important. Hopefully, your teacher will give you lots of opportunities to practise the skills you need in order to do well in this section of the course. That's another reason why we've included the suggested practicals in each unit – so that you can use these experiments to help you get better at planning experiments, collecting data and making conclusions based on the data.

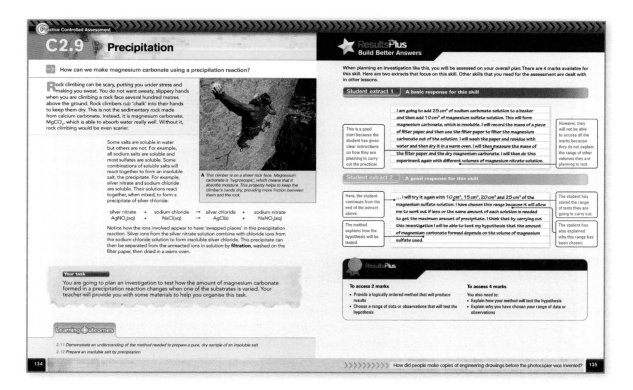

If you've had a flick through this textbook before reading this section, then you'll notice that each of the suggested practicals has a double–page spread in the book and that each of these concentrates on a different area of practical skills. You may do some of these suggested practicals as 'practice' Controlled Assessments to help you prepare for your real Controlled Assessment. Obviously in your real Controlled Assessment you will have a different investigation.

What do you have to do in a Controlled Assessment?

As far as possible, the idea of the Controlled Assessment task is to try and assess how well you have picked up key practical skills during your science course. This means that we have split up each Controlled Assessment into three distinct sections: Part A (planning), Part B (observing) and Part C (concluding).

Planning. In the planning section, you will be given a prediction (or, as scientists often call it, a hypothesis) to test. You'll also be given some information about the sort of experiment you should be doing to test the prediction. Remember, the Controlled Assessment will often be based on one of the suggested practicals that are in the specification – so you should have a good idea of what the experiment that you're trying to plan is like!

For GCSE Additional Science and for the separate sciences, you'll be asked to come up with your own hypothesis or idea to test.

Observing. In this section, you will carry out your plan and collect some data. You'll also research some secondary evidence. Your teacher may give you a pre-prepared plan to follow, rather than using your own. Don't worry if you think your practical skills are a bit basic, or if you know that you work very slowly and carefully when you do a practical. No–one is going to be standing over you to watch exactly what you're doing! We can assess how well you do an experiment by looking at the data that you get.

Concluding. Here, there are some key skills we're looking for: how well you present your data and draw relevant graphs; how you process information in order to make a conclusion; and, finally, whether you think the evidence that you have gathered during your practical supports the prediction or hypothesis you were aiming to test. Your conclusions will need to be based on your data and the secondary evidence that you collected.

It should be pretty clear that the Observing and Concluding sections go together – you can't analyse data if you've not collected any! But the Planning section is separate, so you may end up doing a Planning task based on one practical and the Observing and Concluding tasks on a different practical.

When will you do a Controlled Assessment?

That's up to your teacher – he or she will give you some notice that you'll be taking one of these tasks, so you'll have time to prepare. Doing well in a Controlled Assessment relies on you having picked up a certain number of skills in planning and analysing – so you'll probably be at least halfway through your GCSE course before you do your first piece of assessed work.

How will the real Controlled Assessment work?

It's possible that you'll do a practice before you do the real thing – or that you may do more than one real piece (because your teacher only has to submit your best mark – so if you do more than one, your best mark counts).

As you know, the Controlled Assessment is split into three parts and it's possible that you'll do the different parts at different times. Indeed, it's possible that you'll do the first part (the planning section) from a different Controlled Assessment task than the other two parts.

Why are practicals important?

One of the key things that marks science out as different to other school subjects is that it is practical.

It's important to realise that scientists do practical work not as a way to liven things up, but because science cannot progress without scientists making observations, collecting data and then using that information to back up (or disprove) a theory or hypothesis.

The philosopher Democritus is credited as first coming up with the idea that all substances are made of small particles called atoms, and that atoms cannot be broken up (in Greek, *atomos* means *indivisible*). Although modern atomic theory bases itself on this two-and-a-half-thousand-year-old idea, modern scientists have shown – practically – that this is not quite true. Indeed, harnessing the energy released by splitting the atom was one of the defining moments of the 20th century and, although humans haven't always used this technology for good, this is a source of energy on which we can rely as fossil fuels begin to be used up.

Throughout history, important discoveries in science have been linked to practical work and observations. By doing practical work and collecting data, scientists are able to put forward theories to explain the data produced. Sometimes, of course, the theory comes first; but the practical work is always there to help give evidence to show the theory is correct.

To think of examples from each science, Harvey could never have proposed the theory of blood circulation in humans without having performed experiments and dissections; Humfrey Davy would never have discovered and isolated so many chemical elements without using electricity to break up substances by electrolysis; and where would Newton's theory of gravity have been without the simple – if unplanned – experiment of dropping an apple on his head?

When designing this GCSE course, one aim has been to encourage your teachers to give you better access to practical work than has often been the case in the past. We think that practical work is very important for your enjoyment of science and, because it's also a great way to learn, we've tried to design our GCSE courses to put practical work back into them.

Part A – Planning

Your teacher will issue you with a student brief. This student brief will give you a bit of background, particularly telling you something about the experiment. It might be that the experiment you're doing is similar to one you've already encountered in your course. For GCSE Science, you'll also be given a prediction or a hypothesis to test.

The planning task is under limited control – this means that you can discuss ideas with people in the class and do some other work as preparation. It's possible that you'll spend some time in class with your teacher discussing ideas which may be important to help you plan. Otherwise, you may be asked to do some research for homework to help you prepare for the planning section. You may be given some prompt questions – some questions for you to think about as you do your preparation.

Some key things for you to consider will be:

- **The apparatus you need to do the experiment**. Think about what you are trying to measure and which apparatus will do this best – and be prepared to justify your choice of equipment. Remember that it's often better to provide a labelled diagram of apparatus rather than a list!

- **How many readings to take**. Here, you want to consider the range of readings that you wish to take as well as whether you want to repeat any readings. Again, be prepared to justify the decisions you come to.

- **How to control factors in the experiment**. As well as thinking about how you're going to change the factor you're investigating, you also need to think about how you're going to make sure that other factors stay the same – in other words, how you make sure that the experiment is a fair test.

- **Any relevant safety factors**. Try to make sure that these are relevant to the experiment you're actually doing and not just general things like "tie your hair back".

- **The scientific theory** which can be used to justify your prediction.

The writing-up of your plan should take place in lesson time – depending on the length of your science lessons, you may be given more than one lesson, but the work will be collected in from you at the end of each lesson.

There are slightly different ways in which the planning exercise will be given to you. You'll already have had the briefing document and your teacher may feel, especially if you've been given some prompt questions, that you've got enough information to write your plan. Some teachers may issue you with the prompt questions again, in written form, just to remind you. Alternatively, you may be given something that looks a bit more like a question paper, with the prompt questions laid out one by one, with blank space left for you to answer each one. Whichever form you get, remember the key questions above and try to write a logical, concise plan that covers all of those points.

Hopefully, you'll get a chance to carry out your plan, so make sure that it's clear to follow and has all the necessary detail.

What makes a good plan?

Let's say that the hypothesis you're given is:

"The greater the mass of fuel burnt, the greater the rise of temperature in some water heated by the burning fuel."

The following is one possible plan to test the hypothesis. It is not a bad plan, but can you see where it can be improved?

"I'm going to measure the rise in temperature when I burn 3 different fuels. I will put the burning fuel underneath a beaker of water and use the burning fuel to heat the water. I'll use a thermometer to measure the temperature of the water before the experiment and again once I have burnt the fuel. I'll measure out 100cm³ of water each time using the scale on the side of the beaker. I'll use a balance to measure the mass of the fuel burnt. The water being heated in the experiment might get quite hot, so I will have to take care when I handle the beaker containing hot water."

The list of apparatus is not very clear – what is being used to hold the fuel, for example? Also, no justification is given about the apparatus named – especially with regard to how well they make the measurements you need. So, is a beaker really a good piece of apparatus for measuring volume – rather than a measuring cylinder? And how accurate are the thermometer and the weighing balance?

Some factors are controlled – the volume of water is mentioned – but others are not mentioned e.g. using the same beaker each time. If someone else was doing this experiment, they'd need to know when to stop the experiment – this could be when a set mass of fuel has been burned or when there is a set temperature rise in the water. This is an important variable.

There's also no indication of some practical details to improve how well the experiment works e.g. stirring the water as it is heated, or stopping the fuel from evaporating.

It's clear how many fuels are being used – three – but is this a wide enough range of fuels to provide evidence for the hypothesis? There's also no mention of doing the experiment more than once in order to check that the results are correct and can be averaged.

Finally, there is a good, relevant safety feature mentioned; although other risks – especially of the burning fuel – have not been mentioned.

Part B – Observations

In most cases, your Part B task will follow the Planning task, and you will perform the experiment that you have just planned. However, your teacher may give you a new plan to carry out at this point if you have problems with the plan you wrote.

In some cases you may just complete the planning task because your teacher may choose to take the Planning task from one area of the course e.g. from Biology; and your Observations task from another
e.g. Chemistry.

Remember, though, that the Part C (Conclusions) task will be based on the Observations task – so it is important that you collect a good set of data.

In many ways, the Observations task is the most straightforward – you will be following your plan or worksheet for the practical work you have to carry out. There are some things for you to consider when you carry out the practical and collect data or observations.

How many readings will you take for each experimental trial you do?

- If you're taking more than one reading, how are you going to check to see that your results each time are concordant? (Concordant is the scientific way of saying that the results agree with each other).

- You may also need to decide on a range of readings to take. For instance, if the plan tells you to set up a circuit and to vary the voltage in the circuit and measure the current, then you may have to decide on a range of voltages to use e.g. 2V, 4V, 6V, 8V, 10V and 12V.

How are you going to record your observations?

- Think here of whether you need any units in the data you are recording, and how many significant figures or decimal places you need to record.

The Observations task is also under limited control. This means that you may work in a group to collect data; and also discuss the range and number of readings you will take within the group. If you are working in a group, it's important that you take turns to do the stages in the experiment – don't rely on one person to do all the practical work and one person to do all the recording!

The final part of the Observations task will be for you to collect secondary evidence. Secondary evidence is often results from a similar experiment or an experiment testing a similar hypothesis. You'll be given some guidance on where to look for your secondary evidence.

Once you have collected data and observations, your teacher will probably ask you to hand these in. This is because you will need this information in the next part of the Controlled Assessment task and so it's very important that it is not lost, or taken home and accidentally left there!

Part C – Conclusions

In order to do the Conclusions task of the Controlled Assessment, you'll have to have completed the Observations task. Don't worry if you've been away when the Observations task was done – your teacher should be able to provide you with some results taken by other students in your class.

This part of the task is under high control. This means that you will be expected to work entirely by yourself. So if you were working in a group in the Observations task, make sure that the data you collected is written in a form that you can understand. It's a good idea to make sure that, if you worked in a group of two for Observations, that you make two copies of the data you collected.

Just as in the Planning task in Part A, there will be a series of prompt questions to help you with this part of the Controlled Assessment. Again, there will be different formats in which these questions may be given to you – if you are in doubt, ask your teacher.

Don't forget the secondary evidence you researched in Part B. You'll need to use this information, along with the data you have collected in Part B, in order to complete Part C.

The key areas for you to concentrate on as you look at the data or observations you have collected are:

- **How you are going to process and present your results**. Most of the time, you'll want to put your results into a table, so think about column headings and units as you draw your table. With many experiments, you'll have data that you can plot to make a graph – but is it best to have a bar chart or a line graph?

- **How well the data you have collected supports the prediction or hypothesis**. You need to be able to use your knowledge of some scientific principles in order to justify the conclusion you make here.

- **An evaluation of the experiment that you did**. Here, you need to think about two things: how the set-up of the experiment worked well and allowed you to collect data of a high quality; and secondly, if there were some areas where the set-up was not helpful. Try to think about how easy the experiment was to do, but also think about how well the equipment you used worked and how good the results you collected were. When you come to think about things which were not good about the experiment, come up with some suggestions about things that you might like to change in order to make it easier to collect results and to make the results more accurate. Lastly, consider what effects any weaknesses in the experiment had on the data you collected and, therefore, how confident you are that your conclusion is correct.

- Don't forget to use the **secondary data** together with your own data when making your final conclusion. And don't forget to use your scientific knowledge to explain why you have come to this conclusion.

Of the three parts of the Controlled Assessment, Part C is probably the most challenging – mostly because you're doing it by yourself. Remember that you can practise the skills needed to succeed in this part of the Controlled Assessment – use your textbook wisely, as it will have lots of advice on tackling the different aspects of the practical work.

And don't forget that you can have more than one go at any part of the Controlled Assessment – it's your best mark for each part of the task that counts!

Biology 2
The components of life

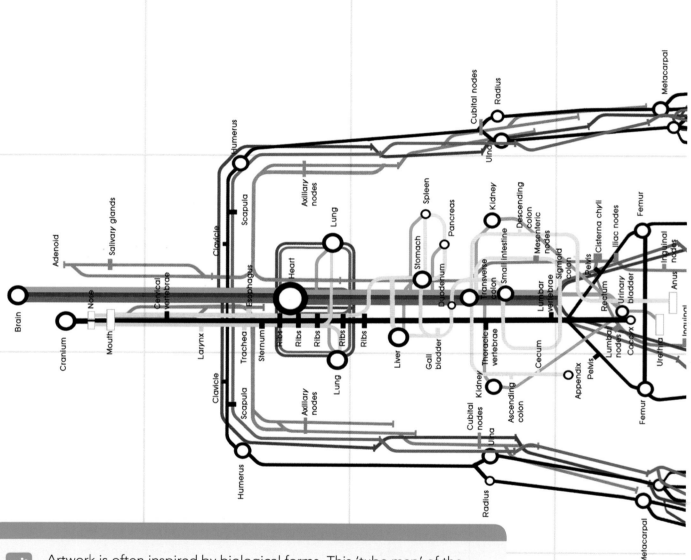

Artwork is often inspired by biological forms. This 'tube map' of the body was created by Dutch artist Sam Loman. It shows some of the main organ systems in the body.

In this unit you will find out about some of the components needed for life, in humans as well as other organisms. The unit will take a look at the basic building blocks of all life before considering some of the processes and systems that are found in plants and animals. You will also consider how the scientific study of the processes that occur within organisms has allowed the development of new technologies.

Index 1
Index 2
Index 3
Index 4
Index 5

Patella
Popliteal nodes
Fibula
Tibia
Index 5
Index 4
Index 3
Metatarsal
Index 2
Index 1

Inguinal nodes
Popliteal nodes
Patella
Tibia
Fibula
Metatarsal
Index 1
Index 2
Index 3
Index 4
Index 5

Index 5
Index 4
Index 3
Index 2
Index 1

Arterial
CNS
Digestive
Lymphatic
Musculoskeletal
Respiratory
Urinary
Venous

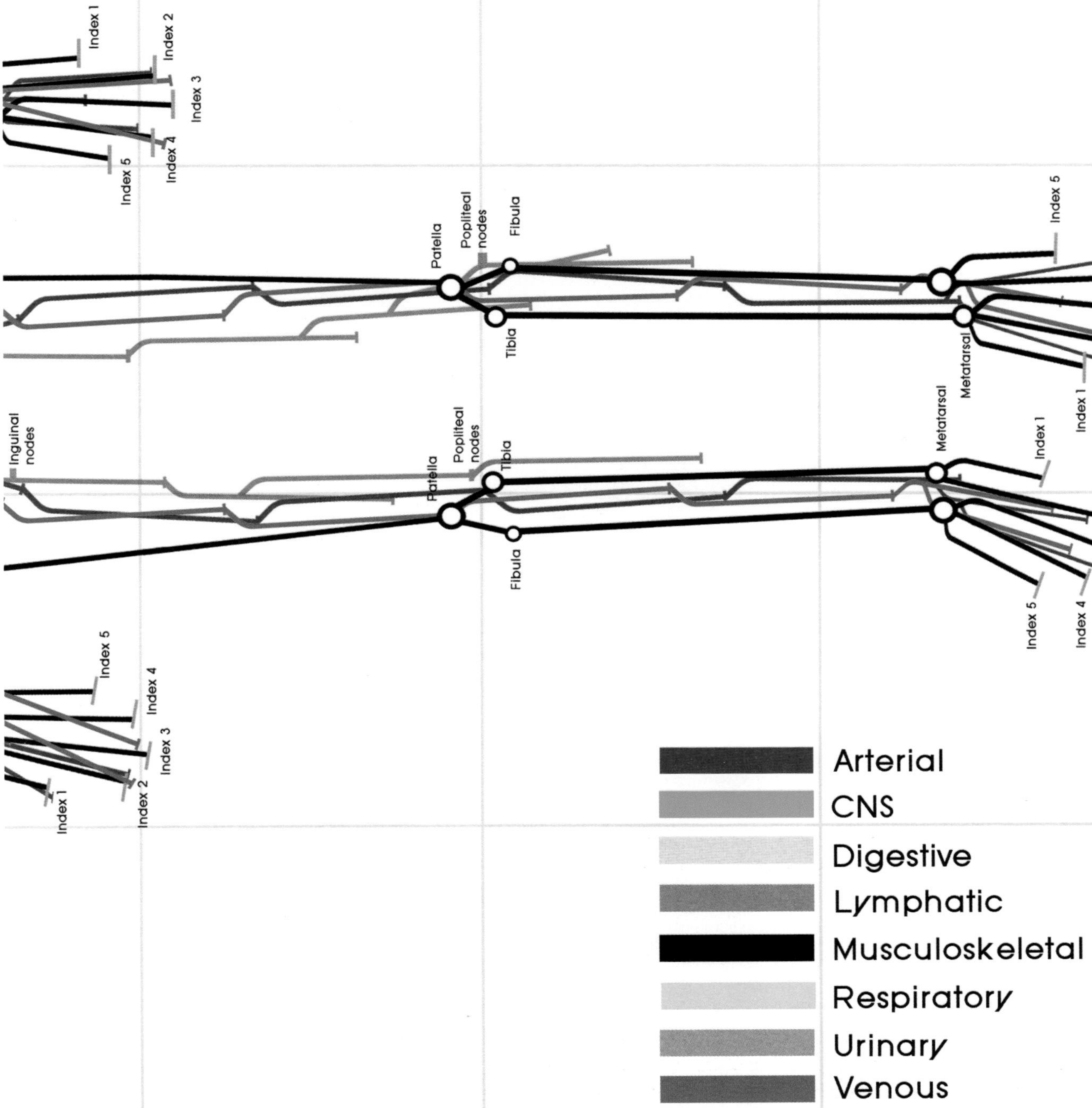

© Transport for body

27

> **What are the differences between plant cells and animal cells?**

For centuries people thought disease was caused by bad air or being cursed. Now we can use light microscopes to see some of the microorganisms that make us ill.

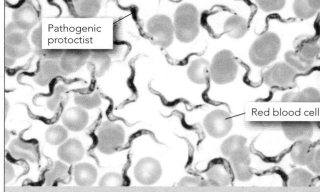

A *A tiny protoctist that causes sleeping sickness. (Magnification × 400)*

In **light microscopes**, light passes through a specimen and then through magnifying lenses so you see the object much bigger than it really is.

Microscopes are used to learn about the structure and the function of **cells**, of which organisms are made.

> **1** Why are microscopes so useful?
>
> **2** What is the purpose of a lens in a microscope?

Plant and animal cells have some features in common:

- **Cell membrane** – separates the contents of the cell and its surroundings. It controls the movement of substances like oxygen, glucose and carbon dioxide into and out of the cell.
- **Cytoplasm** – where many of the chemical reactions needed to carry out life processes take place. It also contains **organelles** (tiny structures that carry out specific jobs).
- **Nucleus** – an organelle that contains **DNA**, which is the genetic material. The nucleus also controls all the activities of the cell.
- **Mitochondria** – organelles in which respiration occurs (oxygen and glucose react to release energy needed by the cell). They are very tiny and cannot be seen easily through a light microscope at low magnification.

> **3** What is the function of the cell membrane?
>
> **4** Why can't you see mitochondria in the light micrograph of a cell?

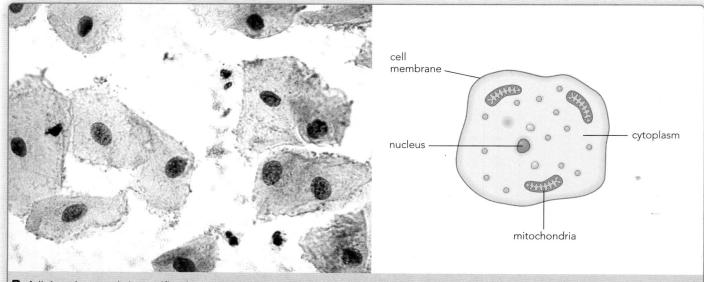

B *A light micrograph (magnification × 2900) and a diagram of a simple animal cell.*

Plant cells

Plant cells also have some other structures:

- **Cell wall** – made of tough **cellulose** to support the cell and allow it to keep its shape.
- **Large vacuole** – a space in the cytoplasm that is filled with cell sap and helps to support the plant by keeping the cells rigid.
- **Chloroplasts** – organelles that contain **chlorophyll**, a green substance that absorbs light energy used in photosynthesis.

Watch Out!

Make sure you don't confuse cell membranes and cell walls. All cells have cell membranes on the outside of the cytoplasm. Only plant cells have cell walls.

Skills spotlight

Sometimes scientists get their inspiration from the work of other scientists. In 1831 Robert Brown (1773–1858) wrote a paper showing that he had found organelles, which he called nuclei, inside the cells of an orchid plant. Suggest what hypothesis Matthias Schleiden (1804–1881) then tested, having read Brown's paper.

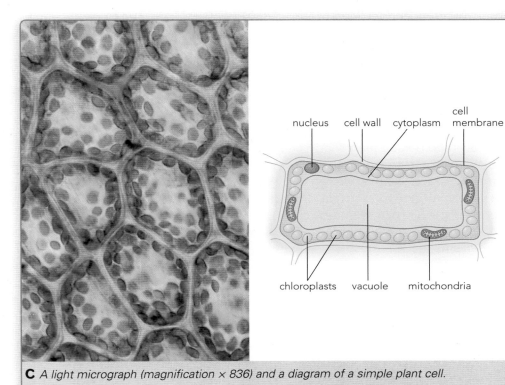

C *A light micrograph (magnification × 836) and a diagram of a simple plant cell.*

5 Why did our knowledge of the structures inside cells only develop after microscopes were invented?

6 a What are the similarities and differences in structure between a typical animal cell and a plant cell?
b Why are some of these features only found in plant cells?

7 List the structures visible in an animal cell and a plant cell using a light microscope, then make a table to show the functions of all the structures.

Learning Outcomes

1.2 Describe the function of the components of a plant cell including chloroplast, large vacuole, cell wall, cell membrane, mitochondria, cytoplasm and nucleus

1.3 Describe the function of the components of an animal cell including cell membrane, mitochondria, cytoplasm and nucleus

1.4 Describe how plant and animal cells can be studied in greater detail with a light microscope

HSW 2 Interpret data, using creative thought, to provide evidence for testing ideas and developing theories

B2.2 Inside bacteria

What are bacterial cells like?

Antoni van Leeuwenhoek (1632–1723) used a microscope he had made to look at plaque he scraped off his teeth. He discovered it contained what he called *tiny animalcules*. They were what we now call bacteria.

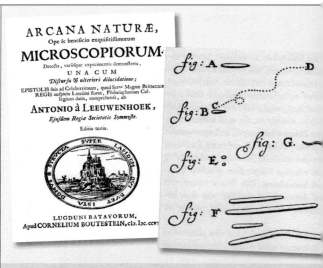

A *Bacteria as Leeuwenhoek drew them in 1683, in a letter to the Royal Society.*

Microscopes have improved greatly in the last 350 years. The best modern light microscopes can magnify specimens more than 1500 times. An average person magnified this much would be over 2.5 km tall! With a magnification like this, we can observe very small structures inside cells. For example, we can see inside bacterial cells. Bacteria are single-celled organisms that are much smaller than plant or animal cells.

1 What are the parts labelled A, B, C and D on the electron micrograph of a bacterium in Figure B?

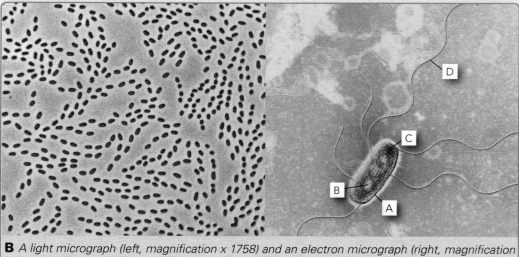

B *A light micrograph (left, magnification x 1758) and an electron micrograph (right, magnification x 2626) of bacteria. The difference in both magnification and clarity is easy to see.*

2 What are the differences between the two types of DNA found in bacterial cells?

In the 1930s the **electron microscope** was invented. This uses a beam of electrons to magnify specimens up to about 2 000 000 times. This would make an average person 3333 km tall! Electron microscopes also produce very clear images and have helped us to discover much more about the detailed structure of cells. For example, light microscopes show us that bacteria do not have nuclei. Electron microscopes show us that bacterial cells contain two types of DNA. **Chromosomal DNA** is a giant loop of DNA containing most of the genetic material. It is not neatly packaged like plant or human chromosomes. **Plasmid DNA** comes in small loops and carries extra information.

Bacterial cells

Using information from the electron microscope we know that bacterial cells have a cell wall that is different to the cell wall of plants. It does a similar job but it is more flexible and not made of cellulose.

Some bacteria also have **flagella** on the outside of the cell. These are long, whip-like structures that bacteria can use to move themselves along.

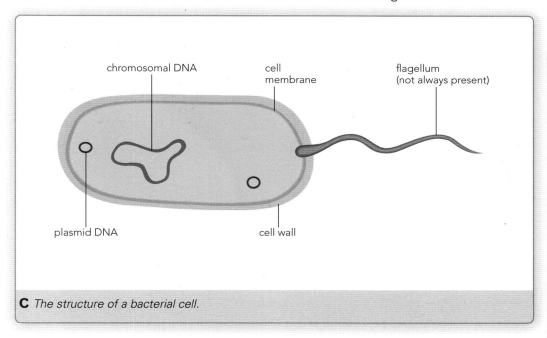

C *The structure of a bacterial cell.*

3 What are the advantages of the electron microscope compared to the light microscope?

4 a What are the magnifications of a good light microscope and an electron microscope?
b How much more powerful is the electron microscope?

5 a A specimen appears 1 mm in length under the light microscope at a magnification of × 1500. What is its actual length?
b What length would the specimen in part **a** appear under the maximum magnification of an electron microscope?

6 Draw and label a diagram to show the structure of a bacterial cell and the functions of each part.

7 Explain the importance of the development of light and electron microscopes.

ResultsPlus
Watch Out!

Remember that bacterial cells have a cell wall, only one circular chromosome but no nucleus.

Maths skills

The length of the magnified object = the length of the object × the **magnification**
And by rearranging this formula we know that:

$$\frac{\text{the length of the object}}{} = \frac{\text{the length of the magnified object}}{\text{the magnification}}$$

The first paragraph of page 30 tells us that an average person magnified 1500 times would be 2500 m tall. From this we can work out the size of an average person:

$$\frac{\text{height of an average person}}{} = \frac{2500}{1500}$$
$$= 1.666 \text{ m}$$
$$= 1.7 \text{ m (to 1 d.p.)}$$

We can say that the length of the magnified object and the length of the object are in **direct proportion** as they increase and decrease in the same ratio.

Skills spotlight

What did Van Leeuwenhoek think bacteria were? How have his ideas been tested and changed since 1683?

Learning Outcomes

1.1 Describe the function of the components of a bacterial cell including chromosomal DNA, plasmid DNA, flagella and cell wall

1.5 Demonstrate an understanding of how changes in microscope technology have enabled us to see cells with more clarity and detail than in the past, including simple magnification calculations

HSW **14** Explain how uncertainties in scientific knowledge and scientific ideas change over time and the role of the scientific community in validating these changes.

What is DNA?

On the 17th of July 1918, the Russian secret police shot the Russian royal family, including the Tsar's youngest daughter, Grand Duchess Anastasia. A woman, called Anna Anderson, later claimed that she was Anastasia and had escaped. Anderson died in 1984. In 1994 DNA testing on some of her tissue showed that she was not Anastasia. Another body was later found and shown to be Anastasia's from DNA testing completed in 2009.

A *DNA analysis showed Anna Anderson was really a Polish woman called Franziska Schanzkowska.*

The **chromosomes** in the nuclei of your cells contain your genetic information. They are made of a chemical called **DNA**. Sections of the DNA molecule called **genes** contain instructions for specific proteins that are used in your body.

1 What is a gene? ?

One gene contains the instructions for one protein. Often several genes work together to produce what is needed for a single feature. For example, your eye colour is the result of proteins made by several different genes. So is the size and shape of your nose, and many other features of your body. Some genes also carry the instructions for the proteins that control the chemical processes in your body.

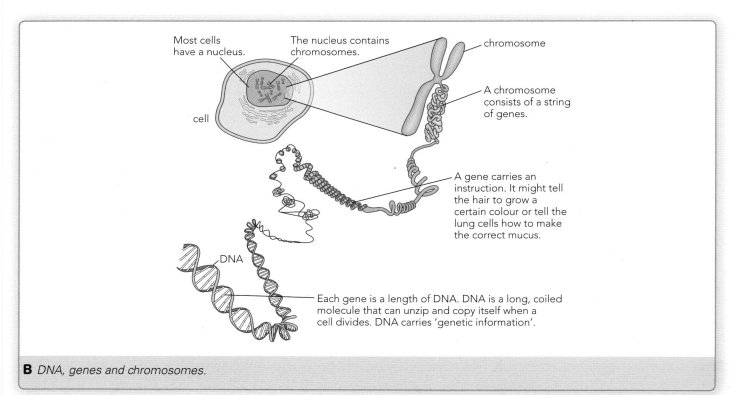

Most cells have a nucleus.

The nucleus contains chromosomes.

chromosome

cell

A chromosome consists of a string of genes.

A gene carries an instruction. It might tell the hair to grow a certain colour or tell the lung cells how to make the correct mucus.

DNA

Each gene is a length of DNA. DNA is a long, coiled molecule that can unzip and copy itself when a cell divides. DNA carries 'genetic information'.

B *DNA, genes and chromosomes.*

The structure of DNA

There are two strands in a molecule of DNA that are coiled together to form a spiral known as a **double helix**. The two strands are linked together at regular intervals by chemicals called **bases**.

The bases always pair up in the same way because of the complementary (matching) shape of the molecules. **Adenine** (A) pairs with **thymine** (T) and **cytosine** (C) pairs with **guanine** (G). The matching bases are known as **complementary base pairs**. They are joined together by weak **hydrogen bonds**. The order of the bases in DNA contains the information needed to form the proteins for body cells. We each have a slightly different order of bases in our genes and this makes us all slightly different.

2 Suggest why a different order of bases in a gene for hair colour in two people may make them have different coloured hair.

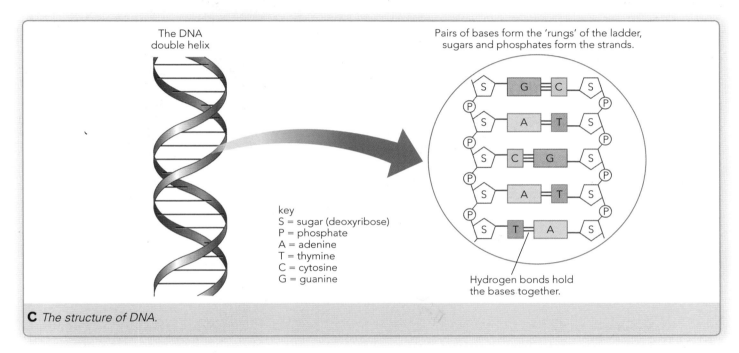

The DNA double helix

Pairs of bases form the 'rungs' of the ladder, sugars and phosphates form the strands.

key
S = sugar (deoxyribose)
P = phosphate
A = adenine
T = thymine
C = cytosine
G = guanine

Hydrogen bonds hold the bases together.

C *The structure of DNA.*

3 a What shape is the DNA molecule?
b What are the four bases found in DNA?
c A DNA molecule is made up of two strands. How are they joined together?

4 Why does adenine always bond with thymine and never with cytosine?

5 Why is DNA so important?

6 The actor Clint Eastwood, who is now in his seventies, publicly thanked his mother 'for her genes'. Explain what a gene is and what he meant.

Skills spotlight

If you commit a crime, your DNA is kept on a database. Some people argue that there should be a national DNA database with all of us on it. Who do you think should be consulted about a decision like this?

Learning Outcomes

1.6 Recall that a gene is a section of a molecule of DNA and that it codes for a specific protein

1.7 Describe a DNA molecule as:
a two strands coiled to form a double helix
b strands linked by a series of complementary base pairs joined together by weak hydrogen bonds:
   i adenine (A) with thymine (T) **ii** cytosine (C) with guanine (G)

HSW 13 Explain how and why decisions that raise ethical issues about uses of science and technology are made

B2.4 Extracting DNA

 Can you extract DNA from common foods?

In 2009 an American man was imprisoned for selling cheap, frozen catfish as expensive fresh grouper, red snapper and flounder. The scam was discovered by extracting the DNA from the fish, testing it and comparing it to the DNA from known species.

The genetic information carried in DNA is different for each species. This means DNA can be used to identify different organisms. For example, scientists have shown that some chicken meat is contaminated with beef because they have discovered cow DNA in the chicken breasts! To do this the first step is to extract DNA from the cells. Some cells contain more DNA than others.

A *The label can claim this fish is anything from catfish to cod – but the DNA doesn't lie!*

Your task

You are going to plan an investigation that will allow you to find out whether changing the method of DNA extraction improves the DNA yield. Your teacher will provide you with some materials to help you organise this task.

Learning Outcomes

1.8 Investigate how to extract DNA from cells

When planning an investigation like this, you will be assessed on your *overall plan.* There are 4 marks available for this skill. Here are two extracts focusing on this skill. Other skills that you need for the assessment are dealt with in other lessons.

Student extract 1 — A basic response for this skill

Give detailed instructions because ideally someone else should be able to carry out the practical in the same way and get the same results.

> I will put the mixture in a hot water bath for a while and then cool it down again. To make the mixture I am going to mix detergent with salty water and then mix this with some peas that have been mashed up. I will then put the mixture in a blender, blend it, filter it and add an enzyme. I will then pour cold liquid down the side of the test tube and the DNA will float on the top. I am going to test lots of different vegetables like this.

Your instructions should be in a logical order so that they are easy to follow.

It would be better to explain what sort of vegetables and how many.

Student extract 2 — A good response for this skill

This student has missed part of the plan out – they should have explained how this method will allow them to test their hypothesis.

> I am going to measure out 50 g of thawed frozen peas and mash them up. I will then mix them with salty water which includes 3 g of salt dissolved in 90 cm³ of water with 10 cm³ of detergent added. The mixture will then be left in a water bath at 60 °C for 15 minutes and then cooled in an ice bath for 5 minutes. I will then filter the mixture and add 2 drops of a protease enzyme. Finally I will pour 10 cm³ of iced ethanol down the side of the tube and DNA will float to the surface of the tube. I will measure the DNA layer with a ruler to see what thickness it is. I will do this experiment again for soya beans, sweetcorn and cress.

Stating the quantities makes it easy for someone else to follow the method.

This time the student has clearly explained the range of vegetables to be tested.

ResultsPlus

To access 2 marks

- Provide a logically ordered method that will produce results
- Choose a range of data or observations that will test the hypothesis

To access 4 marks

You also need to:
- Explain how your method will test the hypothesis
- Explain why you have chosen your range of data or observations

DNA discovery

What is the human genome project?

Each year scientific Nobel Prizes are awarded for chemistry, physics and medicine. Each prize-winner gets a solid gold medal and about £1 000 000 (although because scientists collaborate this is often shared).

A *In 1962 James Watson and Francis Crick, along with Maurice Wilkins, were awarded a Nobel Prize for discovering the structure of DNA.*

B *This X-ray photograph of DNA, taken by Franklin, shows that DNA is a double helix. However you need to be an expert to interpret the image.*

In London in the 1950s, Maurice Wilkins (1916–2004) and Rosalind Franklin (1920–1958) were studying the structure of DNA using X-rays. Franklin directed beams of X-rays at purified DNA and used photos to record how the molecule scattered the X-rays. From the patterns (see Figure B) she could work out how the groups of atoms in the DNA molecule were arranged.

In Cambridge, James Watson (1928–) and Francis Crick (1916–2004) were trying to build a 3D molecular model of DNA using the data from a number of other scientists. They used X-ray results from the London team including some of Franklin's best X-ray images that Wilkins had shown them without Franklin's permission. It was the detail in her images that gave Watson and Crick the clues they needed to build their double helix model.

However, when Watson and Crick published their paper in 1953, their only reference to the London team was a footnote. Eventually the contribution made by Franklin and Wilkins became clear. In 1962 James Watson, Francis Crick and Maurice Wilkins won a Nobel Prize for discovering the structure of DNA. Sadly, Rosalind Franklin died aged just 37, 4 years before the Nobel Prize was awarded.

1 Who discovered the structure of DNA? **(?)**

2 How did the London team and the Cambridge team differ in the way that they tried to work out the structure of DNA?

Understanding the human genome **H**

The **Human Genome Project (HGP)** was an international effort that involved scientists working in 18 different countries and sharing the data they collected using the latest developments in IT to help them.

In 2003, after 13 years of work, the HGP was complete. Scientists had worked out the sequence (order) of the 3 billion base pairs that make up the human **genome**. Although each human being has a unique DNA sequence, the project shows that everyone has at least 99.9% of their DNA in common!

H 3 What is the difference between a gene and a genome? **(?)**

Knowing the sequence of the human genome has many implications for science and medicine. It is being used to develop:

- improved testing for genetic disorders, to discover if people are carrying a faulty allele
- new ways of finding genes that may increase the risk of certain diseases, e.g. Alzheimer's, heart disease
- new treatments and cures for disorders, e.g. gene therapy, where scientists try to replace or mend faulty genes that cause the disorder (see Figure C)
- new ways of looking at changes in the genome over time – this shows us how humans have evolved and the evolutionary relationships between different species
- personalised medicines – these are medicines that work with a particular genotype and target diseases more effectively, with fewer side effects.

C This boy, Salsabil Abu Said, used to suffer from a disease called SCID. His immune system was not working and he had to be shielded from contact with people. In 2002, he was cured using gene therapy. However, things are not always so successful. In 1999, Jesse Gelsinger died while taking part in a gene therapy experiment.

ResultsPlus
Watch Out!

A genome is all the genes in an individual. Do not confuse genes and alleles. A gene is part of a chromosome and alleles are varieties of one gene.

Skills spotlight

The HGP involved a huge amount of data that had to be collected and analysed. Explain how this was achieved in a short period of time and justify this approach.

H 4 a What was the human genome project?
b How did the attitude of the scientists involved in the HGP differ from those who worked on the structure of DNA?

H 5 The HGP analysed the DNA from five different individuals. Why might this limit the knowledge we have and how could we solve this problem?

H 6 What ethical issues does the HGP raise? You could divide your answer into the following areas: testing for genetic disease, gene therapy and personalised medicine.

7 Describe the roles played by the different scientists in the discovery of the structure of DNA.

Learning Outcomes

1.9 Explain how the structure of DNA was discovered, including the roles of the scientists Watson, Crick, Franklin and Wilkins

H 1.10 Demonstrate an understanding of the implications of sequencing the human genome (Human Genome Project) and of the collaboration that took place within this project

HSW **1** Describe the collection and analysis of scientific data

Genetic engineering

 How can we best use genetic engineering?

To show that a genetically engineered characteristic can be passed on from parents to their offspring, the pigs in Figure A were genetically engineered so their trotters glowed green in UV light. When they bred naturally, piglets with green-glowing trotters was born.

A *Genetically engineered pigs – their trotters glow in UV light (top).*

> **1** What is genetic engineering?
>
> **2** Why are enzymes important in genetic engineering?

Scientists can remove a gene from one organism and insert it into the DNA of another organism. This process is called **genetic engineering**. For example, scientists have inserted the gene for human insulin into bacterial plasmid DNA. The genetically modified (GM) bacteria can then make human insulin, which is used by people with diabetes. Organisms like these bacteria are known as **genetically modified organisms** (**GMOs**).

The production of human insulin by genetically modified (GM) bacteria has many advantages. Insulin used to be extracted from dead cattle and pigs but insulin from GM bacteria can now be used by vegans. It also means that the supply of insulin is not affected by animal diseases or the numbers of animals slaughtered for meat. Using bacteria in fermenters, insulin can be made in vast quantities and can be made more cheaply. The one slight disadvantage is that as bacteria produce the insulin slightly differently there are minute differences and it does not suit everyone.

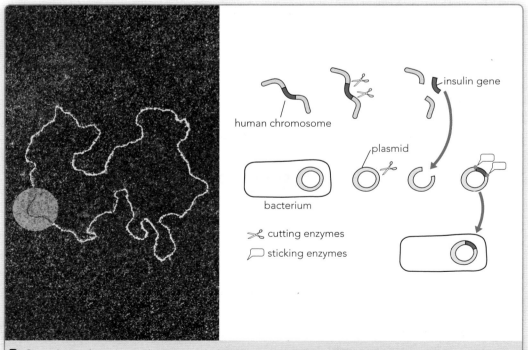

B *Genetic engineering – the green circle in the photo on the left shows the extra bit of DNA that has been added to this bacterial plasmid; the diagram shows the process of adding a human insulin gene to a bacterial plasmid.*

Plants and animals can also be genetically engineered. For example, **golden rice** plants are normal rice plants that have had two extra genes inserted so that they make **beta-carotene** in the grains. Beta-carotene is needed by humans to produce vitamin A. A lack of vitamin A can cause death because the immune system does not work properly. It can also cause blindness.

However, some people are concerned that the GM rice will crossbreed with wild rice plants and contaminate the wild rice DNA. Others worry that eating GM organisms might harm people (although there is no evidence of this). Some people say the levels of beta-carotene in golden rice are not high enough to make much difference to children's health. GMOs can also be expensive to buy and some are made so that they do not produce fertile seed (meaning farmers have to buy new seed each year).

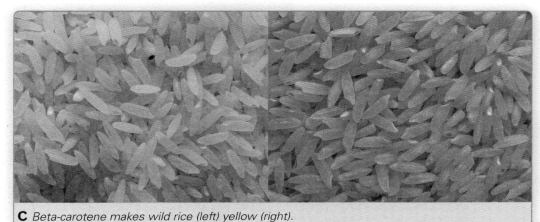

C *Beta-carotene makes wild rice (left) yellow (right).*

Scientists have added genes to some plants to make them **herbicide** resistant. This can reduce the amount of crop spraying needed because the farmer can use one much heavier spray of herbicide rather than several smaller doses. There may be drawbacks, including the development of herbicide-resistant weeds as cross-pollination with wild plants takes place. There may also be a loss of **biodiversity** as fewer weeds survive, meaning loss of food and shelter for animals.

>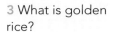
>
> 3 What is golden rice?
>
> 4 How would eating golden rice solve the problems caused by lack of vitamin A?
>
> 5 Make a flowchart to show the steps in the process of genetic engineering.
>
> 6 Make a table to show the advantages and disadvantages of GM golden rice.

>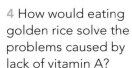
> ## Skills spotlight
> Draw up a table to compare the benefits of GM human insulin compared to insulin taken from cattle and pigs slaughtered for meat.

> 7 a Explain the potential benefits of bacteria engineered to produce human insulin and of herbicide-resistant plants.
> b Using the same examples as in part a, discuss some of the drawbacks of genetic engineering.
>
> 8 Write a piece for a campaign website, either *for* or *against* genetic engineering. You must explain the process and why it is done, and then give your own opinion.

Learning Outcomes

1.11 Demonstrate an understanding of the process of genetic engineering, including the removal of a gene from the DNA of one organism and the insertion of that gene into the DNA of another organism

1.12 Discuss the advantages and disadvantages of genetic engineering to produce GM organisms, including:
   a beta-carotene in golden rice to reduce vitamin A deficiency in humans
   b the production of human insulin by genetically modified bacteria
   c the production of herbicide-resistant crop plants

HSW *12* Describe the use of contemporary science and technological developments and their benefits, drawbacks and risks

B2.7   **Mitosis and meiosis**

What are mitosis and meiosis?

Cells usually send signals to each other to stop dividing when there are enough of them. Cancer cells don't respond – they just keep dividing. Scientists are investigating the use of GM bacteria to transport proteins into cancer cells to kill the cancer cells.

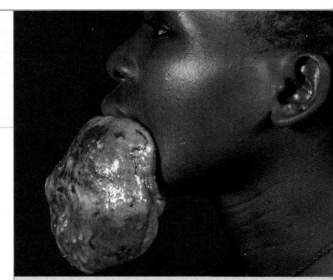

A *A cancer is caused by uncontrolled cell division.*

Mitosis

Most of the cells in your body contain a nucleus, and most nuclei contain two copies of each chromosome. This makes them **diploid** cells. Human diploid cells contain two sets of 23 chromosomes. Body cells (all cells except sperm and egg cells) are diploid.

1 Explain what is meant by a diploid cell.

2 Using Figure B, describe as fully as you can what happens during mitosis.

To make more body cells, during **growth** or to replace damaged cells, body cells divide using a process called **mitosis**. This process begins with the chromosomes making copies of themselves in a process called **DNA replication**. The copies of the chromosomes separate and the cell then divides. This produces two **daughter cells**, which are genetically identical to each other and to the **parent cell**.

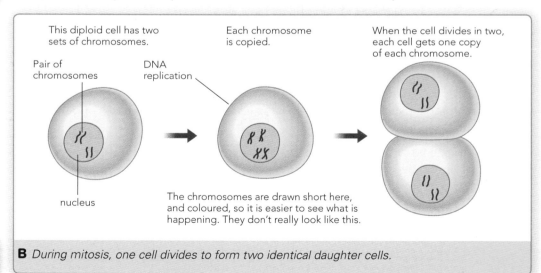

B *During mitosis, one cell divides to form two identical daughter cells.*

3 How many parents does an individual have in organisms that reproduce by:
a asexual reproduction
b sexual reproduction?

Different kinds of reproduction

Mitosis can produce new, complete individuals in a process called **asexual reproduction**. Bacteria cells often do this by splitting in half. Some plants reproduce asexually by making new plantlets attached to the parent plant, which then split off to grow on their own.

In many organisms a different process called **sexual reproduction** produces new individuals. This requires two sex cells or **gametes**. These are different to body cells in that they only contain one set of chromosomes, so these are **haploid** cells. The two gametes fuse during **fertilisation** and produce a diploid cell called a **zygote**. The zygote develops into a ball of cells (an **embryo**).

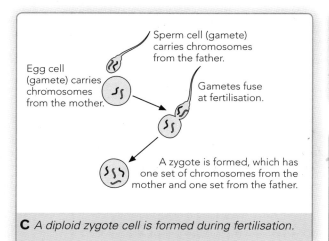

Egg cell (gamete) carries chromosomes from the mother.

Sperm cell (gamete) carries chromosomes from the father.

Gametes fuse at fertilisation.

A zygote is formed, which has one set of chromosomes from the mother and one set from the father.

C *A diploid zygote cell is formed during fertilisation.*

Gamete formation

A different kind of cell division called **meiosis** is needed to produce haploid gametes. Meiosis also starts with DNA replication, but this is then followed by *two* cell divisions. The first separates the two sets of chromosomes and the second separates the copies of each chromosome. This produces four haploid daughter cells, each containing one set of chromosomes.

In a diploid cell, each chromosome in a pair contains the same genes but may have different versions of those genes (**alleles**). So the chromosomes in a pair are slightly different. In meiosis, these slightly different chromosomes are split between the daughter cells in a random way to produce gametes that are genetically different from each other.

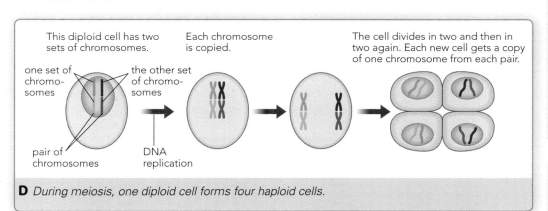

This diploid cell has two sets of chromosomes.

one set of chromosomes

the other set of chromosomes

pair of chromosomes

DNA replication

Each chromosome is copied.

The cell divides in two and then in two again. Each new cell gets a copy of one chromosome from each pair.

D *During meiosis, one diploid cell forms four haploid cells.*

6 Are all the body's cells genetically identical or not? Explain your answer.

4 What is the process in which a zygote develops into an embryo?

ResultsPlus
Watch Out!

Students often confuse the spelling of mitosis and meiosis. The words look similar but they describe very different processes. Remember: Mi2osis for *two* sets of chromosomes, Me1osis for *one* set of chromosomes.

Skills spotlight

Scientists present information in diagrams to help explain the relationship between processes. Draw a flowchart of a human life cycle and mark the presence of diploid and haploid cells in the cycle. Identify when mitosis and meiosis occur to produce these cells.

5 Draw up a table to summarise the similarities and differences between mitosis and meiosis.

Learning Outcomes

1.13 Describe the division of a cell by mitosis as the production of two daughter cells, each with identical sets of chromosomes in the nucleus as the parent cell, and that this results in the formation of two genetically identical diploid body cells

1.14 Recall that mitosis occurs during growth, repair, and asexual reproduction

1.15 Recall that, at fertilisation, haploid gametes combine to form a diploid zygote

1.16 Describe the division of a cell by meiosis as the production of four daughter cells, each with half the number of chromosomes, and that this results in the formation of genetically different haploid gametes

HSW *11* Present information using scientific conventions and symbols

Would it be a good idea to make copies of your favourite pet?

B2.8 Clones

 How are organisms cloned?

In 2008 Bernann McKinney paid £25000 to have her pet dog Booger cloned. She had to sell her home so that she could pay for the process, in which skin cells from Booger developed to produce five puppy clones.

A *Booger's clones*

B *Leaf cuttings taken from a plant produce new plant clones.*

Clones are individuals that are genetically identical. Making plant clones can be easy – you start with a bit of leaf, stem or root from the original plant. The plant cells divide and produce new cells, which grow into a clone of the original plant. This is an example of asexual reproduction.

Benefits, drawbacks and risks of cloning mammals

It is much more difficult to clone animals than plants – it is not possible to make a whole new animal from an arm or leg. The first clone produced in a lab of a vertebrate was a frog, and was made in 1952. However, the first cloned large animal (a sheep called Dolly) wasn't produced until 1996 because of difficulties with the process.

Unfortunately, very few embryos produced during animal cloning develop successfully. Dolly was the only lamb produced after 237 cloning attempts. Dolly also seemed to grow older much more quickly than normal and died young. Scientists aren't sure if her health problems were caused by being a clone or by chance.

Cloning is useful to make a genetically identical copy of an adult organism that has desirable characteristics. For example, bulls whose sperm produces high-quality calves are valuable and so are worth cloning. Cloning is also used to produce copies of individuals with a genetically engineered trait, such as cows engineered to produce human insulin in their milk, because this will guarantee that all the offspring will have the engineered trait.

1 What is a clone?

2 Are new plants from cuttings produced by meiosis or mitosis? Explain your answer.

3 Give a benefit of cloning animals.

4 What are some of the drawbacks and risks of cloning animals?

C *Cloned puppies of South Korea's best sniffer dog are being tested to see if they are as successful as their parent.*

How to clone a mammal

There are different ways of producing cloned animals, but currently the most successful method uses a process called nuclear transfer. The nucleus of a body cell of the animal to be cloned is transferred into an **enucleated** egg cell (one that has had its original nucleus removed). The cell is then stimulated to start dividing to form an embryo. This is **implanted** into the **uterus** (womb) of a **surrogate mother** who is a different individual to the parent. Here it will grow and develop into a new individual.

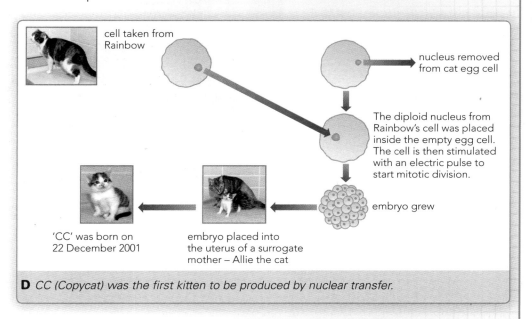

cell taken from Rainbow

nucleus removed from cat egg cell

The diploid nucleus from Rainbow's cell was placed inside the empty egg cell. The cell is then stimulated with an electric pulse to start mitotic division.

embryo grew

'CC' was born on 22 December 2001

embryo placed into the uterus of a surrogate mother – Allie the cat

D *CC (Copycat) was the first kitten to be produced by nuclear transfer.*

Skills spotlight

New scientific developments usually have drawbacks and risks as well as benefits. What are the benefits, drawbacks and risks of using cloned bulls for breeding programmes?

ResultsPlus
Watch Out!

Students often panic if they see an unfamiliar example in the exam. You will be expected to apply what you know about cloning to examples of different animals and plants. Remember that asexual reproduction produces clones.

(H) 5 Review Figure D and explain the roles of Rainbow and Allie in the production of the kitten called CC, and explain which adult CC is a clone of.

(H) 6 Describe as fully as you can the stages in producing a cloned mammal.

7 An internet forum has this question on it: 'I have produced many of my African violet plants from a single plant by taking cuttings. My wonderful pet dog, Bruno, is getting old and I've heard that scientists can do similar things to dogs. Does this work and what are the risks?' Write an answer to this question.

Learning Outcomes

1.17 Recall that cloning is an example of asexual reproduction that produces genetically identical copies

(H) 1.18 Demonstrate an understanding of the stages in the production of cloned mammals, including:
- *a* removal of diploid nucleus from a body cell
- *b* enucleation of egg cell
- *c* insertion of diploid nucleus into enucleated egg cell
- *d* stimulation of the diploid nucleus to divide by mitosis
- *e* implantation into surrogate mammals

1.19 Demonstrate an understanding of the advantages, disadvantages and risks of cloning mammals

HSW 12 Describe the benefits, drawbacks and risks of using new scientific and technological developments

How are stem cells useful?

The girl in Figure A is Cady, who was killed in a car crash in 2002. After her death, Dr Panayiotis Zavos took some of her cells and stored them. He wants to use them to create a cloned embryo and bring to life a person who is genetically identical to Cady.

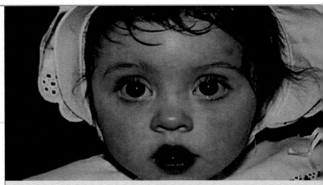

A *Cady – a girl that Dr Panayiotis Zavos wants to bring back to life.*

B *A human embryo produced from a cloned adult cell.*

In most countries trying to produce cloned humans is illegal. However, in the UK scientists can produce cloned human embryos for research but these embryos must be killed after 14 days.

There is a lot of interest in human embryos because when they are a few days old the cells in them are **embryonic stem cells**.

When a **stem cell** divides it can produce more stem cells or it can produce other types of cells that are specialised or **differentiated** (e.g. neurones, muscle cells, skin cells). Once a cell has differentiated it cannot turn into another type of cell. Stem cells in differentiated body tissues (**adult stem cells**) can only differentiate into a few types of cell. However, embryonic stem cells can develop into almost every type of human cell.

One way of collecting embryonic stem cells is to use leftover embryos created for couples having fertility treatment. However, when the embryonic stem cells are extracted, the embryo is killed, so this procedure is controversial. That's why scientists are interested in using adult cells to make cloned embryos.

1 Explain the meanings of these terms:
a differentiate
b stem cell.

2 Name four different kinds of differentiated cell in the body.

3 Describe two sources of stem cells.

C *Some people think that because human embryos are potential people, destroying them is murder.*

Stem cell treatments

Since the 1960s, adult stem cells from bone marrow have been used to treat leukaemia (a cancer of certain white blood cells). The patient's white blood cells are destroyed and adult stem cells from someone else are put into the patient. These cells then multiply and produce new, healthy white blood cells. This is known as a bone marrow transplant. The treatment does not always work because the body destroys cells from other people if they are too different.

This problem may be solved using cloning. If a cloned embryo is created using a skin cell from a person with leukaemia, the embryonic stem cells can be taken and used to produce the cells that make white blood cells. These will survive inside the patient because their body will recognise them as their own cells.

Since embryonic stem cells can turn into nearly every type of human cell, these cells could be used to treat many more problems than adult stem cells could. However, this technique could also be used illegally by people like Dr Zavos to produce human clones.

Scientists have started to investigate ways of turning differentiated body cells into stem cells by reprogramming them. If this works, it will help avoid the ethical problem of using embryos. But further research is still needed to make any treatment with stem cells safe because if they are injected into a body they may produce the wrong kind of cell or even create cancer cells.

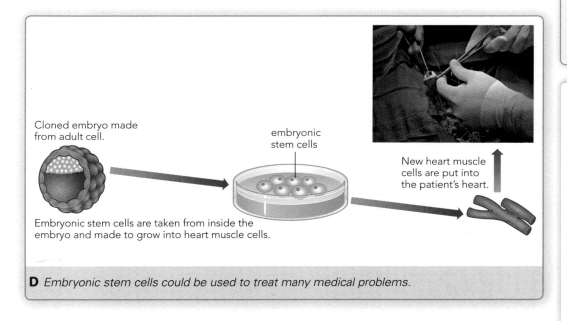

Cloned embryo made from adult cell.

embryonic stem cells

New heart muscle cells are put into the patient's heart.

Embryonic stem cells are taken from inside the embryo and made to grow into heart muscle cells.

D *Embryonic stem cells could be used to treat many medical problems.*

ResultsPlus
Watch Out!

Students often confuse adult and embryonic stem cells. Make sure you learn the difference and are clear in your answers which sort of stem cell you are writing about. Embryonic stem cells can develop into almost any type of cell.

Skills spotlight

Science alone does not provide a method for making decisions about ethical issues. Even though scientists can use adult cells to make cloned embryos to get stem cells, is it right to use these embryos and destroy them?

4 Draw up a table to show the benefits, drawbacks and risks of using embryonic stem cells and adult stem cells to treat human disorders.

5 Compare and contrast the benefits and drawbacks of using embryonic stem cells and adult stem cells in research and treatment.

Learning Outcomes

1.19 Demonstrate an understanding of the advantages, disadvantages and risks of cloning mammals

1.20 Recall that stem cells in the embryo can differentiate into all other types of cells, but that cells lose this ability as the animal matures

1.21 Demonstrate an understanding of the advantages, disadvantages and risks arising from adult and embryonic stem cell research

HSW 13 Explain how and why decisions that raise ethical issues about uses of science and technology are made

How are proteins made?

Haemophilia is a disease in which blood does not clot properly. This can be caused by a lack of a blood protein called Factor VIII. In the 1980s scientists examined the structure of Factor VIII. A protein's structure is produced by a code in the order of bases in the gene. Knowing the structure of the protein allowed the scientists to work back to what the gene must look like. They found the gene in 1984.

A Today, the gene for Factor VIII can be produced in the milk of these genetically engineered pigs.

The order of bases on a DNA strand forms the **genetic code**. There are only four bases (adenine, A; cytosine, C; guanine, G; thymine, T) but they can be arranged in any sequence, for example ATTAGCG.

There are 20 different **amino acids** in human proteins. A cell uses the sequence of bases in DNA to synthesise (build) chains of these amino acids. These chains then form proteins. This process is called **protein synthesis**.

Each amino acid is identified by a different group of bases. A specific order of bases in DNA produces a specific order of amino acids in the chain and so produces a particular protein.

1 What are proteins made of?

2 What controls the sequence of amino acids that are joined together to make a specific protein?

3 Why do different parts of DNA produce different proteins?

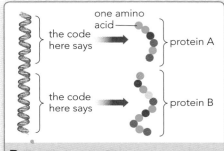

B Translating the DNA code into proteins.

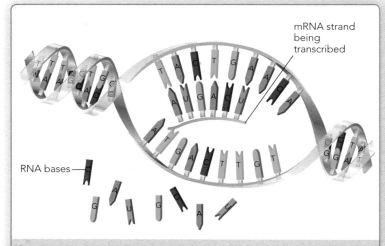

C Transcription of DNA to produce mRNA.

Transcription

The process of making a protein takes place in two stages. The first is **transcription**, which takes place inside the nucleus. Here, the DNA in a gene unzips by breaking the weak hydrogen bonds between bases in the double helix. One strand of the gene is used as a template. Bases that are complementary to this strand link together opposite it, forming a molecule of **messenger RNA** (**mRNA**).

RNA is very similar to DNA but only has one strand (not two) and has a base called **uracil** (**U**) instead of thymine. mRNA is small enough to move out of the nucleus into the cell's **cytoplasm**.

Translation

In the cytoplasm, the mRNA attaches to a small structure called a **ribosome**. The ribosome moves from one end of the mRNA strand to the other, decoding the bases in groups of three, known as **base triplets** or **codons**. Each amino acid is attached to a **transfer RNA (tRNA)** molecule, and each tRNA molecule has a triplet of bases. The triplet of bases controls which amino acid is attached.

As the ribosome moves to the next codon on the mRNA, the tRNA with complementary bases lines up with the codon. The tRNA releases the amino acid that it was carrying which joins on to the growing amino acid chain. The tRNA is released as the ribosome moves on to the next codon, where the next tRNA with its amino acid lines up. This continues until the mRNA strand is completely decoded. The chain of amino acids is called a **polypeptide**. The polypeptide then twists and folds and may link up with other polypeptides, to become a protein.

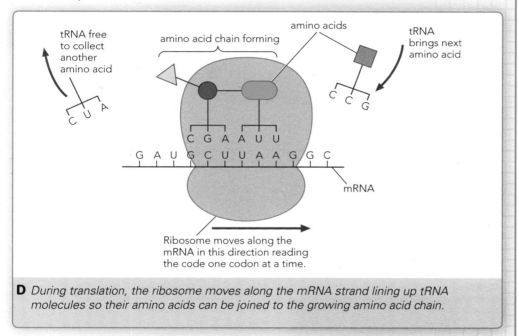

D During translation, the ribosome moves along the mRNA strand lining up tRNA molecules so their amino acids can be joined to the growing amino acid chain.

ResultsPlus
Watch Out!

To avoid muddling the words *transcription* (in the nucleus) and *translation* (in the cytoplasm). Remember transcribing is the copying of something exactly and translating is changing it into something different (like another language), in this case changing from base codes into amino acid sequence in a protein.

H **4** Describe the differences between translation and transcription.

Skills spotlight

Scientists use models to help understand and explain what they see. A car factory, where a whole car is put together in a sequence from pieces using a plan, could be used as a model for protein synthesis. What are the strengths and weaknesses of this model?

5 Describe as fully as you can how different proteins are made from the genetic code. Include bases, amino acids and proteins in your answer.

Learning Outcomes

1.22 Describe how the order of bases in a section of DNA decides the order of amino acids in the protein

H **1.23** Demonstrate an understanding of the stages of protein synthesis, including transcription and translation:
  a the production of complementary mRNA strand in the nucleus
  b the attachment of the mRNA to the ribosome
  c the coding by triplets of bases (codons) in the mRNA for specific amino acids
  d the transfer of amino acids to the ribosome by tRNA
  e the linking of amino acids to form polypeptides

HSW **3** Describe how phenomena are explained using scientific models

How do mutations happen?

Stumpy the duck has a rare condition called polymelia. This means that the genes in his body made four legs, not the usual two.

A *Stumpy*

> **1** Find the names of three proteins on this page.
>
> **2** What is the link between the type of protein that keratin is and where it is found in the body?
>
> **3** Explain why the shapes formed by the amino acid chains in haemoglobin are important.

The number and order of the amino acids in the chain is specific for each protein. For example the protein insulin, which helps cells take in glucose, has a small chain of 51 amino acids in an order that no other protein has. Haemoglobin is a much larger, complex protein found in red blood cells. It carries oxygen and is made of four linked chains, two of which have 141 amino acids and two have 146 amino acids.

The order of the amino acids causes the chain to fold up in a particular way and gives the protein a specific 3D shape. Some proteins such as keratin found in human hair and nails, form long, strong fibrous molecules. Other proteins such as insulin, haemoglobin and enzymes, have a round 'globular' (blobby) shape. This helps them move around inside cells and the rest of the body easily. The shape of enzymes is particularly important in the way they work. Each enzyme has a unique shape caused by its amino acid sequence. An enzyme can only work in one kind of reaction because of its shape. We say that each enzyme is specific to the reaction.

B *Keratin is the main protein in stags' antlers.*

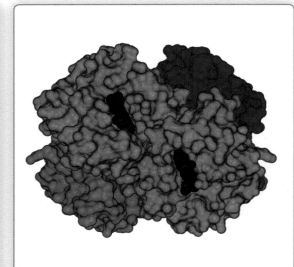

C *The 3D shapes formed by the different chains of amino acids in haemoglobin means it can carry oxygen.*

The effect of mutations

A **mutation** is a change in the sequence of bases in the genetic code. Some changes in the code have no effect on the amino acid sequence produced, so the protein shape is not affected. Others, such as the sickle-cell mutation in the gene that produces haemoglobin, replace one amino acid in the chain with another. This can cause the protein to fold up in a different way and have a different shape, which will affect the way it works.

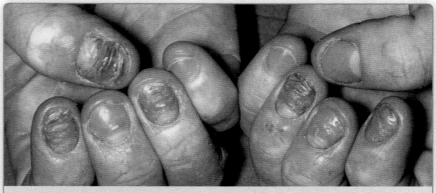

D *A mutation in a keratin gene can cause misshapen nails.*

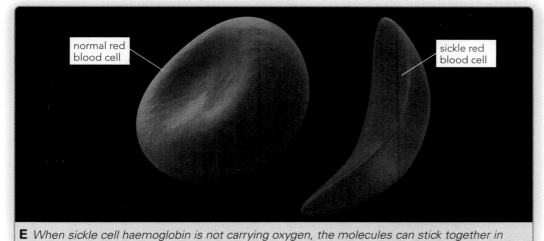

E *When sickle cell haemoglobin is not carrying oxygen, the molecules can stick together in long fibres, making the red blood cells pointy and sickle-shaped. These cells easily get stuck in small blood vessels.*

ResultsPlus
Watch Out!

Do not forget that a mutation in a gene can produce a new allele. The normal allele for the haemoglobin gene can mutate to become the sickle-cell allele.

Mutations may be harmful or may have no effect, and some can be beneficial to the organism. For example, some mutations in bacteria make them resistant to the effects of antibiotics. Bacteria that have the mutation survive when that antibiotic is used, but bacteria without the mutation die.

Skills spotlight

Theories can help us explain what we see. Explain how the theory that a protein is coded for by the order of bases in DNA can help explain the effect of the sickle-cell mutation.

4 Describe the effect of the sickle-cell mutation on cell shape.

5 Explain why the sickle-cell mutation is a problem.

6 A mutation in the gene for keratin causes brittle hair. Explain as fully as you can why it does this. Include DNA, amino acids and proteins in your answer.

Learning Outcomes

1.24 Describe each protein as having its own specific number and sequence of amino acids, resulting in different shaped molecules that have different functions, including enzymes

1.25 Demonstrate an understanding of how gene mutations change the DNA base sequence and that mutations can be harmful, beneficial or neither

HSW 3 Describe how phenomena are explained using scientific theories and ideas

How can an enzyme change the colour of your skin?

B2.12 Enigma

B2.12 Enzymes

What do enzymes do?

The colour of your hair and skin is produced by a protein called melanin. An enzyme called tyrosinase is needed to make melanin. A few people have a mutation in the gene for tyrosinase, so they no longer make the enzyme and therefore cannot make melanin.

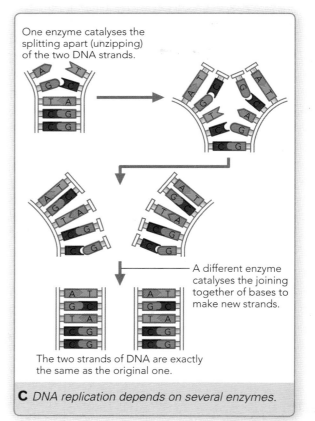

A *This boy does not have dark skin like his father because he has a mutation in the gene for the enzyme tyrosinase*

What are enzymes?

There are thousands of chemical reactions going on in the body at the same time. To work well and quickly, each reaction is controlled by one of a particular group of proteins called **enzymes**. A substance that helps a chemical reaction go faster without itself being changed by the reaction is called a **catalyst**. Without enzymes, these reactions might still happen but at too slow a rate for cells to do all they need to stay alive.

Some enzymes help a large substance to break apart into smaller molecules (such as in **digestion**) and others help smaller chemicals join together to make larger ones (synthesis).

Enzymes inside cells

During DNA replication in mitosis or meiosis, the DNA double helix is unwound and the weak hydrogen bonds separating the two strands are separated by one particular enzyme (rather like unzipping a zip). As new bases line up along each half, so that complementary base pairs match, a different enzyme joins them together. This makes two complete and identical DNA molecules. The enzymes are unchanged so they can repeat their action wherever it is needed in the DNA.

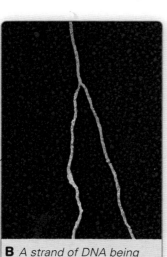

B *A strand of DNA being replicated to make two new, identical strands.*

> 1 What is a catalyst?
>
> 2 Suggest why enzymes are sometimes called 'biological catalysts'.
>
> 3 Explain why the body needs enzymes.

One enzyme catalyses the splitting apart (unzipping) of the two DNA strands.

A different enzyme catalyses the joining together of bases to make new strands.

The two strands of DNA are exactly the same as the original one.

C *DNA replication depends on several enzymes.*

During protein synthesis, when a protein is built from amino acids using the order of bases in the DNA code as a guide, there are many different reactions. Each of these reactions is catalysed by a different enzyme. For example, as a protein is built the reaction that joins one amino acid to another is catalysed by a specific enzyme.

Enzymes outside cells

Food molecules, such as carbohydrates, proteins and fats, are much too large to pass across the cell membranes of the gut wall and into the blood. They need to be broken down first in a process called digestion. Different enzymes are released into the mouth, stomach and small intestine to help digest different food molecules into smaller ones that can be absorbed into the cells.

Microorganisms and fungi also release digestive enzymes, but because they don't have a gut, they grow on and through the food they are digesting. After the enzymes have digested the food, the small molecules are absorbed through the microorganism's cell walls. Some of these enzymes are now used in laundry detergents to help digest food and other large molecules that stain clothes.

D *The strawberries on the right are being digested by the enzymes released by the mould fungus growing on them.*

4 Give an example of where an enzyme catalyses the breakdown of a large molecule inside a cell.

5 Give an example of where an enzyme catalyses the formation of a large molecule from smaller ones inside a cell.

Skills spotlight

The way scientific applications develop is often influenced by need. Since enzymes were first used in laundry detergents, consumers have asked for better cleaning at lower temperatures. Explain how consumer demand can affect the scientific research that is carried out.

6 Why do all organisms need digestive enzymes?

7 Small molecules often dissolve more easily in water than larger ones. Explain as fully as you can why enzymes are used in laundry detergents. Include proteins, catalysts and enzymes in your answer.

Learning Outcomes

1.26 Describe enzymes as biological catalysts

1.27 Demonstrate an understanding that enzymes catalyse chemical reactions occurring inside and outside living cells, including:
a DNA replication
b protein synthesis
c digestion

HSW 13 Explain how and why decisions about uses of science and technology are made

B2.13 Enzymes and temperature

How does temperature affect enzymes?

Frozen peas can contain higher levels of nutrients, such as vitamins, than fresh peas sold in shops. This is because cell processes in the peas start breaking down the nutrients as soon as the peas are picked. Freezing the peas slows these processes down to a very slow rate.

The processes in pea cells that break down nutrients are like many processes in cells – they are controlled by **enzymes**. Enzymes help to speed up the rate of reactions. If the enzymes are affected in any way, the rate of reaction usually slows down. We can use this idea to test how different factors can affect enzymes.

A Within 2 hours of being picked, these peas will be frozen and ready for packaging.

Your task

You are going to plan an investigation that will allow you to find out how the rate of an enzyme-controlled reaction is affected by temperature. Your teacher will provide you with some materials to help you organise this task.

Learning Outcomes

1.32 Investigate the factors that affect enzyme activity

When planning an investigation like this, one of the skills you will be assessed on is your ability to *write a hypothesis.* There are 4 marks available for this skill. Here are two extracts focusing on this skill. Other skills that you need for the assessment are dealt with in other lessons.

Student extract 1 | A basic response for this skill

Use your scientific knowledge to say why you think your hypothesis is correct.

> I think that if you make a reaction hotter then it will *go faster* and I think that reactions which contain enzymes probably work like this just like any other reaction so I am going to get some amylase and put it in a test tube and test what happens when I heat it up.

You just need one sentence which says what you think will happen in your investigation.

You don't need to give any practical detail in this section.

Student extract 2 | A good response for this skill

> I think that changing the temperature of a reaction which is controlled by an enzyme will change the rate of the reaction up to a certain point. After this point I think that the rate of reaction will slow down. I think this because I know that when the temperature of any reaction increases the molecules have more energy and are more likely to collide and react. I also know that enzymes are damaged if they get too hot so I think that if you heat the reaction up too much the enzyme will be damaged, it will not work and the reaction will slow down.

This is a clear hypothesis which will be easy to test.

This student has done a good job of using their own relevant scientific knowledge to justify their hypothesis.

It would be better to use the scientific word 'denatured' here.

ResultsPlus

To access 2 marks
- Provide a hypothesis that is appropriate for most of the task
- Try to justify your hypothesis

To access 4 marks
- Provide a hypothesis that is appropriate for the full scope of the task
- Justify your hypothesis using relevant scientific ideas

How do enzymes work?

Bacteria that grow inside giant columns of ikaite (a kind of rock) in deep water near Greenland contain enzymes that work well at very low temperatures. Using enzymes like these in laundry detergents could mean washing in cold water still gets clothes clean.

A An ikaite column in deep water near Greenland

1 What is the normal temperature of the main human organs?

2 What is meant by optimum temperature?

The molecules that enzymes work on are called **substrate** molecules. Enzymes catalyse the change of substrate molecules into product molecules better at some temperatures than others, so the rate of reaction is faster.

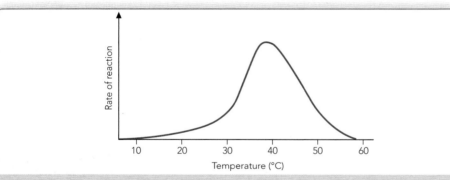

B Many human enzymes have an optimum temperature that is close to the normal temperature of the main organs.

At temperatures much above or below an optimum (best) value, enzymes don't work as well. This is partly why the processes in your body don't work as well when you have a fever.

Enzymes also often work best at an optimum pH. Most enzymes in cells work best at about pH7, but enzymes in the digestive system have to work well at much higher or lower pHs.

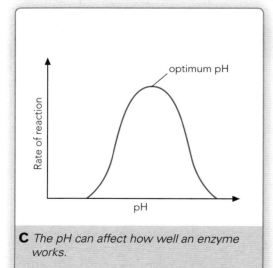

C The pH can affect how well an enzyme works.

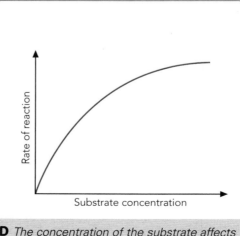

D The concentration of the substrate affects the rate of reaction.

The rate of a reaction catalysed by an enzyme will also increase as the concentration of substrate increases but only up to a point. Beyond that concentration, there is no further change in reaction rate. This is because the enzyme cannot work on the substrate any faster than at this rate. So adding more substrate molecules will make no difference to the rate.

Specific enzymes

Another key feature of enzymes is that each enzyme will only work with a particular substrate, or small group of similar substrates. We say they are highly **specific** for the substrate. This feature is used to name enzymes, for example:

- carbohydrases catalyse the breakdown of carbohydrates
- proteases catalyse the breakdown of proteins.

Explaining the observations

All the substrate molecules for one particular enzyme have the same 3D shape in some part of their molecules. This suggests that shape is important in enzyme action.

Looking at the 3D shapes of enzymes and their substrates shows that the substrate fits neatly into an **active site** in the enzyme. The active site has a different shape in different enzymes. Since the shape of the substrate fits tightly into the hole of the active site, this model of how enzymes work is called the '**lock-and-key' hypothesis**.

Changing the pH or temperature a little changes the shape of the active site so the substrate does not fit as well. Too much change will break the bonds within the enzyme. This can change the shape so much that it **denatures** the enzyme and destroys the active site.

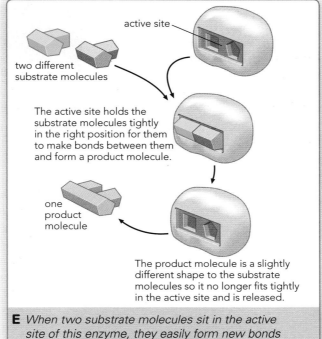

active site

two different substrate molecules

The active site holds the substrate molecules tightly in the right position for them to make bonds between them and form a product molecule.

one product molecule

The product molecule is a slightly different shape to the substrate molecules so it no longer fits tightly in the active site and is released.

E *When two substrate molecules sit in the active site of this enzyme, they easily form new bonds to produce one new product molecule.*

3 Stomach contents have a pH of about 2. Sketch a graph to show the rate of reaction against pH for an enzyme that digests food in the stomach.

4 State three factors that can affect how well enzymes work.

Watch Out!

Students often use the phrase 'enzymes are denatured' without explaining that the high temperature changes the shape of the active site and the substrate cannot bind to it.

Skills spotlight

Scientists interpret data to provide evidence for testing ideas. Explain how the results from the effect of temperature on enzyme activity support the idea of an active site in enzymes.

5 How does the 'lock-and-key' hypothesis explain the specificity of enzymes for their substrates?

6 Describe how the 'lock-and-key' hypothesis explains the synthesis of one molecule from two substrates. Include the active site in your answer.

Learning Outcomes

1.28 Describe the factors affecting enzyme action, including:
   a temperature *b* substrate concentration *c* pH

1.29 Recall that enzymes are highly specific for their substrate

1.30 Demonstrate an understanding of the action of enzymes in terms of the 'lock-and-key' hypothesis

1.31 Descrbe how enzymes can be denatured due to changes in the shape of the active site

HSW *2* Describe how data is used by scientists to provide evidence that increases our scientific understanding

3 Describe how phenomena are explained using scientific theories, models and ideas

How do cells get what they need for aerobic respiration?

Eero Mäntyranta has unusual blood. As a result of an inherited mutation his blood contains higher than normal numbers of red blood cells. This gave him a competitive advantage over other athletes.

A *Eero Mäntyranta, from Finland, won seven Olympic cross-country skiing medals in the 1960s.*

All organisms are made of cells and energy is needed to power the many processes a cell carries out. This energy is released in a series of enzyme-catalysed reactions known collectively as **respiration**. Put simply, respiration is the release of energy from food molecules that acts as fuel for the cell.

> **1** Suggest why farm animals that are not allowed to move around need less food than farm animals that can wander around freely.

Cells that are more active have higher energy requirements. During exercise the muscles contract to cause movement, and this requires a great deal of energy from respiration. Growing and dividing cells also need lots of energy to power the building of new cell materials.

glucose + oxygen ⟶ carbon dioxide + water

The energy from the glucose is released for use in the cell.

B *Aerobic respiration is a complex series of reactions but what happens can be modelled using a word equation.*

In **aerobic respiration** oxygen is used to release energy from molecules such as **glucose**. The glucose and oxygen are converted into carbon dioxide and water and energy is released for use in the cell.

Delivering glucose and oxygen

In humans, the glucose and oxygen needed for respiration are carried around the body and into tissues by blood. As well as carrying glucose and oxygen, blood must also carry waste carbon dioxide away from respiring cells. All these substances move between respiring cells and tiny blood vessels called **capillaries** by a process called **diffusion**. Diffusion occurs when particles of a substance spread out, moving from an area where they are in higher concentration to an area where they are in lower concentration. The particles diffuse down a **concentration gradient**.

The number of particles decreases as you go down a concentration gradient.

diffusion

net movement of particles

higher concentration   lower concentration

C *The green particles are diffusing along a concentration gradient.*

In respiring cells, oxygen and glucose levels fall as they are used up in aerobic respiration. At the same time, carbon dioxide levels in the cells rise.

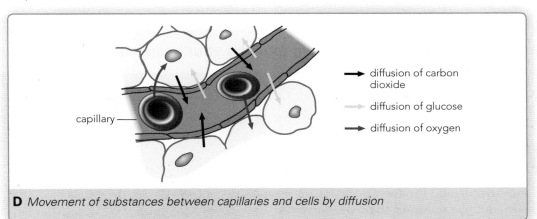

capillary

→ diffusion of carbon dioxide

→ diffusion of glucose

→ diffusion of oxygen

D *Movement of substances between capillaries and cells by diffusion*

Gas exchange

Lung tissue is spongy and full of tiny air sacs (alveoli). These sacs are surrounded by capillaries, and oxygen that enters the body through the lungs moves into the blood by diffusion. Carbon dioxide also leaves the blood by diffusion into the air spaces. As one gas is entering the bloodstream another one is leaving it; this is called **gas exchange**.

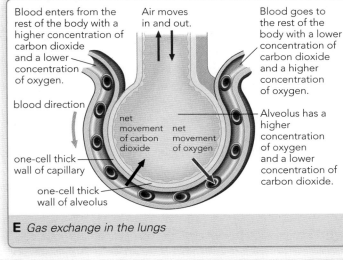

Blood enters from the rest of the body with a higher concentration of carbon dioxide and a lower concentration of oxygen.

Air moves in and out.

Blood goes to the rest of the body with a lower concentration of carbon dioxide and a higher concentration of oxygen.

blood direction

net movement of carbon dioxide

net movement of oxygen

Alveolus has a higher concentration of oxygen and a lower concentration of carbon dioxide.

one-cell thick wall of capillary

one-cell thick wall of alveolus

E *Gas exchange in the lungs*

4 Describe how oxygen enters the blood in the lungs.

5 Glucose and oxygen are needed for aerobic respiration to occur. Which of these two molecules contains the energy that is released in respiration?

6 Describe two processes inside cells that require energy to occur.

7 Suggest why Eero Mäntyranta had a competitive advantage over most other athletes.

2 Explain why oxygen diffuses into respiring cells but carbon dioxide diffuses out of them.

3 Capillaries have very thin walls. Why do you think this is important?

Skills spotlight

Models can help us to understand how processes happen at a molecular level.

a Explain how a word equation is a model for what happens to a molecule of glucose in aerobic respiration.

b Breaking bonds between atoms requires energy and making bonds between atoms releases energy. Use the word equation model and your knowledge of bonds to describe simply how energy is released in aerobic respiration.

ResultsPlus
Watch Out!

Students often lose marks because they think that respiration is the same as breathing and gas exchange. Remember that breathing is to do with lungs, gas exchange occurs at the cell surface and respiration happens inside the cell.

Learning Outcomes

2.1 Recall that respiration is a process used by all living organisms that releases the energy in organic molecules

2.2 Explain how the human circulatory system facilitates respiration including: **a** glucose and oxygen diffuses from capillaries into respiring cells **b** carbon dioxide diffuses from respiring cells into capillaries

2.3 Define diffusion as the movement of particles from an area of high concentration to an area of lower concentration

2.4 Demonstrate an understanding of how aerobic respiration uses oxygen to release energy from glucose and how this process can be modelled using the word equation for aerobic respiration

HSW *3* Describe how phenomena are explained using scientific models

B2.16 Investigating the effects of exercise

What effect does exercise have on your breathing rate and your heart rate?

At very high altitude the air pressure and the oxygen concentration of the air is lower than at sea level. This makes it hard for the cells of your body to get enough oxygen. Even the simplest activity is very hard work. You puff and pant, and your heart races, just with the effort of walking.

A Climbers often use oxygen to help their bodies cope with the conditions high up on a mountain.

Climbers and athletes often wear heart beat monitors to show their heart rates when they are climbing mountains or training. You can also monitor your heart rate by taking your pulse. Your pulse is the surge of blood in your arteries as your heart pumps blood around your body, so by counting your pulse you can measure the number of times your heart beats.

When you exercise your muscles need lots of energy. The energy comes from aerobic respiration which needs oxygen (as well as glucose) and produces carbon dioxide. The oxygen and carbon dioxide enter and leave our bodies when we breathe and they are carried in the blood. This means that exercise can cause a change in heart rates and breathing rates.

B You can take your pulse in your wrist. Don't use your thumb because it has a pulse too.

Your task

You are going to plan an investigation that will allow you to find out more about the effect of exercise on your breathing and heart rate. Your teacher will provide you with some materials to help you organise this task.

Learning Outcomes

2.5 Investigate the effect of exercise on breathing rate and heart rate

Build Better Answers

When planning an investigation like this, one of the skills you will be assessed on is your ability to collect and use *secondary evidence*. There are 2 marks available for this skill. Here are two extracts that focus on this skill. Other skills that you need for the assessment are dealt with in other lessons.

Student extract 1 — A basic response for this skill

This evidence is not relevant to the investigation which is being carried out.

> I found a website and I have written the web address below. It has a table which looked at the effect of caffeine on heart rate. It seems that caffeine speeds up the heart rate. I also found a graph in a book which shows me what resting heart rate should be for people of different ages.

Provide a full web address so that someone else could look up the same information.

It is a good idea to give more than one example if you can – although this one does not give us a clear reference so it would be difficult for someone else to find this secondary evidence.

Student extract 2 — A good response for this skill

This practical is relevant to the investigation so is more useful.

The student has evaluated how trustworthy the website is.

> I found a website and I have written the web address below. It has a table of results from a practical which looked at the effect of exercise on heart rate. The web site is run by a university (detailed below) and the results are properly referenced to the work of another scientist so I think that this is a trustworthy web site. I also found a graph in a book which shows me what happens to breathing rates during exercise but there was no detail about how the experiment was carried out, when it was done or who conducted it so I don't think that this source is quite as valuable.

The student has also evaluated the information from the book.

To access 1 mark

- Collect secondary evidence that is relevant to your hypothesis
- Record your secondary evidence in an appropriate way

To access 2 marks

You also need to comment of the quality of the sources of your secondary evidence.

Anaerobic respiration

What effect does exercise have on your body?

The harder you exercise, the faster your heart beats. Free-divers train to lower their heart rate to reduce the rate at which oxygen is delivered to their tissues. This helps them dive for longer on a lungful of air. Trained free-divers can drop their heart rate to 20 beats per minute during a dive.

A Sara Campbell, from London, holds a world record and has dived to 96 m on one breath.

> **ResultsPlus**
> **Watch Out!**
>
> Remember that higher breathing rate results in the blood getting more oxygen and faster heart rate gets more blood to the cells and that both of them help to remove carbon dioxide faster.

The more active a cell is the more energy it needs so respiration must happen at an increased rate. During exercise the muscles use up oxygen and glucose very quickly, so the blood supply to the muscles must increase.

The amount of blood circulated by the heart depends not only on the heart rate, but also the **stroke volume**. This is the volume of blood pumped out of the heart on each beat. If we multiply heart rate by stroke volume, we get the **cardiac output**, which is the volume of blood circulated by the heart in a given time.

cardiac output = stroke volume × heart rate

Heart rate and stroke volume increase with exercise, so cardiac output also increases. This delivers more blood to respiring tissues faster.

Extra blood is of limited use to the respiring tissues if it doesn't contain enough oxygen. To make sure it does, our breathing rate increases to increase the rate of oxygen uptake in the lungs.

B Muscle contraction needs a constant supply of energy.

1 Describe what happens to heart rate and breathing rate during exercise.

2 a Write an equation for calculating cardiac output.
b At rest, an athlete's heart rate is 60 beats per minute, and stroke volume is 76 cm³. During exercise these increase to 160 beats per minute and 105 cm³. Calculate the increase in cardiac output.

When we carry out intense exercise we cannot supply oxygen to our muscles quickly enough. When this happens an alternative process to break down glucose and release energy that does not rely on oxygen starts to occur alongside aerobic respiration. It is called **anaerobic respiration**.

| glucose | ⟶ | lactic acid |

The energy from the glucose is released for use in the cell.

C Anaerobic respiration can be modelled using a word equation.

Anaerobic respiration releases less energy than aerobic respiration. **Lactic acid** is broken down, using oxygen, into carbon dioxide and water. After exercise increased oxygen is required to break down lactic acid and to release energy for other processes in cells (for example cell repair). This requirement for additional oxygen after exercise is called **excess post-exercise oxygen consumption** (**EPOC**). This used to be called the 'oxygen debt'.

This extra oxygen can be obtained and distributed by maintaining a high breathing rate and a high heart rate which is one of the reasons why they remain relatively high for a few minutes after exercise. The time taken for the pulse rate to return to the normal or resting rate after exercise is the recovery time.

D *Lactic acid produced by anaerobic exercise may cause muscle cramps. After exercise it is broken down, which is one reason for EPOC.*

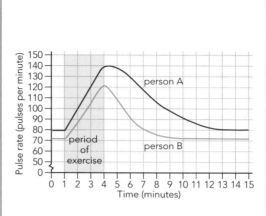

E *Changes in the pulse rates of two people before, during and after exercise.*

3 a Write out the word equation for aerobic respiration and the word equation for anaerobic respiration.
b What is the difference in energy content of the products from aerobic and anaerobic respiration?

4 Explain why the need for increased oxygen does not end when exercise stops.

5 Suggest why it is better to take deep breaths for a few minutes after you have stopped running, instead of shallow ones.

6 The heart muscle needs a supply of energy to power its contractions. During exercise, how will the energy needs of the heart muscle itself change?

7 An old textbook contains the sentence 'The oxygen debt is caused by the body's need for oxygen to break down lactic acid'. Rewrite this sentence for a more up-to-date textbook.

8 It is possible to jog for several hours. Suggest why it is not possible to sprint for several hours.

Skills spotlight

Graphs can be useful to compare sets of data. Which of the people in Figure E has the fastest recovery time? Explain your answer.

Learning Outcomes

2.6 Explain why heart rate and breathing rate increase with exercise

2.7 Calculate heart rate, stroke volume and cardiac output, using the equation
cardiac output = stroke volume x heart rate

2.8 Demonstrate an understanding of why, during vigorous exercise, muscle cells may not receive sufficient oxygen for their energy requirements and so start to respire anaerobically

2.9 Demonstrate an understanding of how anaerobic respiration releases energy from glucose and how this process can be modelled using the word equation for anaerobic respiration

2.10 Recall that the process of anaerobic respiration releases less energy than aerobic respiration

2.11 Describe how a build-up of lactic acid requires extra oxygen to break it down. This is called excess post-exercise oxygen consumption or EPOC (formerly known as oxygen debt)

2.12 Explain why heart rate and breathing rate remain high after exercise

HSW *11* Present information, develop an argument and draw a conclusion, using scientific, technical and mathematical language

>>>>>>>>>>>>>>>>>>>>>>>> How do plants get glucose for respiration if they don't eat it in food?

What is starch, and where does the energy it contains come from?

Marathon runners often change their diet several days before a race, eating large amounts of starchy food like potatoes, rice or pasta. The excess carbohydrate is stored in the form of 'glycogen' and can be broken down when glucose is needed to power their long journey.

A *A lot of starch is found in foods like potatoes and pasta. Plants use starch to store energy.*

> **1** State two differences between the reactions of photosynthesis and aerobic respiration.

Starch is made by joining together thousands of glucose molecules. When starch is broken down in digestion, glucose is released and can be used for respiration. Plant cells also need to respire and so need a supply of glucose. Plants can actually manufacture their own glucose using the simple raw materials of carbon dioxide and water. This process is called **photosynthesis**. Plants can also store the glucose they make as carbohydrates, like starch.

the energy from sunlight

carbon dioxide + water ⟶ glucose + oxygen

B *Photosynthesis can be modelled using a word equation.*

Photosynthesis is a complex series of enzyme-catalysed reactions which occurs inside cell organelles called **chloroplasts**.

Light is the energy source for photosynthesis and is absorbed by a green substance inside chloroplasts called **chlorophyll**. Without chlorophyll, photosynthesis cannot occur. Chlorophyll transfers the light energy into the stored chemical energy in glucose.

> **2 a** Which parts of the variegated leaf in Figure C lack chlorophyll?
> **b** Suggest why variegated plants tend to grow more slowly than normal plants.

C *The leaves of this variegated plant have some areas without chloroplasts. The leaf on the right has been tested for starch using iodine solution which turns blue-black if starch is present. Only the areas of the leaf that contain chlorophyll contain starch.*

Leaf adaptations

The leaf is the main plant organ in which photosynthesis occurs. Leaves have several adaptations for photosynthesis, including the presence of chloroplasts containing chlorophyll to absorb light energy. Leaves are broad and flat to provide a very large surface area which helps them absorb as much light energy as possible.

On the underside of a leaf there are microscopic pores called **stomata** (singular = **stoma**). The stomata open in response to light. These pores allow carbon dioxide in the atmosphere to diffuse into the leaf, to be taken up by photosynthesising cells. Oxygen produced by these cells can also diffuse from the inside of the leaf out to the atmosphere. Plant cells produce water in respiration, some of which evaporates from their surfaces and diffuses out of the leaf.

The air spaces inside the leaf give the cells a large **surface area to volume ratio** for efficient gas exchange.

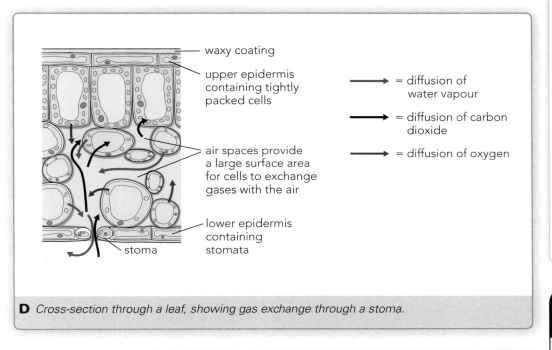

= diffusion of water vapour

= diffusion of carbon dioxide

= diffusion of oxygen

D *Cross-section through a leaf, showing gas exchange through a stoma.*

5 If the stomata on a leaf become blocked why does this cause the leaf to die?

6 Photosynthesis can take place in any plant cells that contain chloroplasts. Explain how leaves are adapted to make them very efficient as sites for photosynthesis.

3 Stomata usually close when it is dark and open when it is light. Why do you think they do it this way round and not the other?

4 If a potato is left in sunlight it will develop a green colour on its surface. Suggest what is responsible for the green colour and why this might be an advantage for a potato.

Watch Out!

Many students think that plants don't respire. All plant cells respire and they have mitochondria just like animal cells.

Learning Outcomes

2.13 Describe how the structure of the leaf is adapted for photosynthesis, including:
   a large surface area
   b contains chlorophyll in chloroplasts to absorb light
   c stomata for gas exchange (carbon dioxide, oxygen and water vapour)

2.14 Demonstrate an understanding of how photosynthesis uses light energy to produce glucose and how this process can be modelled using the word equation for photosynthesis

HSW 3 Describe how phenomena are explained using scientific theories and ideas

B2.19 Factors affecting the rate of photosynthesis

What factors affect the rate of photosynthesis?

Would you buy a mobile phone, a cooking pot and a car made from plants? More and more of the plastics we use are being made from plant materials instead of chemicals from oil. Fortunately plants produce around 243 billion tonnes of dry plant material every year by photosynthesis. If we are careful there will be plenty of plants to eat as well as all the bioplastics we want and need.

Glucose is made by photosynthesis and it is quickly turned into starch in the plant. The starch is then later broken down into sugars which are used in a variety of ways in the plant.

Oxygen is given off as a waste product during photosynthesis. You can't see this oxygen in land plants, but in water plants such as *Elodea* the oxygen given off is seen as a stream of tiny bubbles.

Factors such as light levels and the temperature of the water will affect the rate at which photosynthesis takes place.

A *This car is the first Formula 3 racing car designed and made from sustainable and renewable materials. For example, the wing mirrors contain potato starch.*

Your task

You are going to plan an investigation that will allow you to find out how light levels affect the rate of photosynthesis. Your teacher will provide you with some materials to help you organise this task.

Learning Outcomes

2.16 Investigate how factors, including the effect of light intensity, CO_2 concentration or temperature, affect the rate of photosynthesis

Build Better Answers

When planning an investigation like this, one of the skills you will be assessed on is your ability to *process evidence.* There are 4 marks available for this skill. Here are two extracts focusing on this skill. Other skills that you need for the assessment are dealt with in other lessons.

Student extract 1 — A basic response for this skill

This is not the way to work out a mean – check that you are using the correct mathematical skill.

> The results I collected from my experiment can be processed like this:
> When the lamp was 10 cm away the number of bubbles in a minute were 22, 20 and 19. The middle reading is 20 so I will take this as my mean.
> When the lamp was 20 cm away the readings were 18, 10 and 18. I can add these up and divide by 3 to get a mean of 15.3.
> I can plot a bar chart to show these results.

This result looks like an anomaly. It may be best not to include it in the mean calculation.

All of this information would be better in a table.

Plotting a chart or graph can be a good way to show your results clearly – but a bar chart is not necessarily the best way to display these results.

Student extract 2 — A good response for this skill

Distance of lamp from plant (cm)	Number of bubbles of oxygen-rich gas produced /min			
	Reading 1	Reading 2	Reading 3	Average (mean)
10	20	22	19	20.6
20	18	19	18	18.3
30	14	14	14	14
40	12	10	9	10.3

Calculating a mean average is a good idea here.

The line graph is a good way to show the pattern in the results. This can help when you try to draw a conclusion.

> I then plotted a line graph which shows the results clearly and has all the correct labels and headings. My line graph is shown below.

ResultsPlus

To access 2 marks

- Attempt to process all your collected evidence
- Use appropriate maths skills
- Present your processed evidence in an appropriate way

To access 4 marks

- Process all your evidence in an appropriate way, using maths skills if appropriate
- Present it in a way that allows you to draw a conclusion

Is there a limit to how much photosynthesis rates can be increased?

If a plant has too little water, it wilts. This causes its stomata to close and so it will not photosynthesise well. Plants grown in warm greenhouses lose a lot of water and are more likely to wilt. This problem is overcome for some crop plants by growing them in water – a technique called hydroponics.

To maximise plant growth, the rate of photosynthesis must be maximised as the glucose made in photosynthesis is used to make proteins for growth, as well for respiration.

For photosynthesis to take place there must be a supply of carbon dioxide, water, a suitable temperature and light to provide the energy for the reactions to occur.

A plant grown in dim light with ample carbon dioxide and water will photosynthesise slowly. Increasing the amount of carbon dioxide or water will not increase the rate of photosynthesis. Only increasing the amount of light will increase the rate. In this case, light is the **limiting factor**.

In any process that is affected by several factors, the maximum rate at which the process can occur is controlled by the factor that is in shortest supply. That factor is the limiting factor.

> 1 What factors could limit the rate of photosynthesis in a plant cell?

A *In hydroponics, plants are grown in a constant supply of water.*

Figure B shows the effect of light intensity on the rate of photosynthesis in a plant (other variables were kept constant). Initially an increase in light intensity causes an increase in the rate of photosynthesis. The plant can photosynthesise faster if it is given more light and so light intensity is the limiting factor in these conditions. In the flat part of the graph the rate of photosynthesis is constant even though more light energy is available. One possibility is that carbon dioxide concentration is now the limiting factor. Another limiting factor could be temperature. The enzymes involved in photosynthesis are working as fast as they can but raising the temperature would allow the rate to increase further.

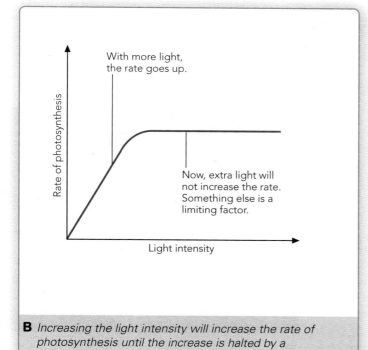

B *Increasing the light intensity will increase the rate of photosynthesis until the increase is halted by a limiting factor.*

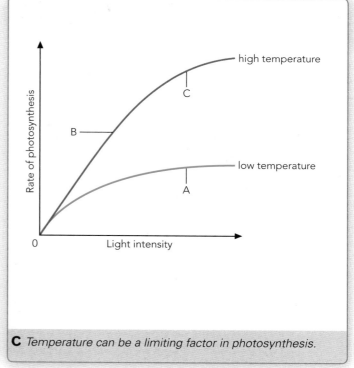

C *Temperature can be a limiting factor in photosynthesis.*

2 a Copy the graph in Figure B and draw in another line to suggest what would happen if the experiment were repeated again with a higher concentration of carbon dioxide.
b If the experiment were repeated at a higher carbon dioxide concentration but the results were exactly the same, what does this tell you about the limiting factor in the flat part of the graph?

3 Explain which factors might be limiting at each stage (A–C) shown in Figure C.

4 Name one limiting factor in a well-lit warm greenhouse and explain how its effect could be reduced.

5 In a forest, some plants grow in the shade and have very dark green leaves. Using this information, suggest what the main limiting factor is for these plants and how they are adapted to survive in these conditions.

6 In a commercial greenhouse, a light automatically switches on at 17:00 in winter and a heater is turned on when the inside temperature drops below 21 °C. Explain both these features of the greenhouse.

Skills spotlight

Explain what features you would look for to see whether a rate of photosynthesis is constant or changing and why plotting a graph makes these things easier to see.

ResultsPlus
Watch Out!

Students sometimes think that photosynthesis happens in the light and respiration in the dark. Respiration happens all the time. If a factor is limiting it does not stop photosynthesis, it means that it will not go any faster.

Learning Outcomes

2.15 Demonstrate an understanding of how limiting factors affect the rate of photosynthesis, including:
   a light intensity
   b CO_2 concentration
   c temperature

HSW 11 Present information using scientific conventions and symbols

>>>>>>>>>>>>>>>>>>>>>>>>>>>> How does water get to the top of a tree? 67

How does water move from the ground up to the leaves of a plant?

Plants normally lose water as water vapour through the stomata in their leaves. If the stomata are all closed, water vapour cannot escape. In some plants water is forced out through special structures in the edges of the leaves. This process is called guttation.

A These droplets were squeezed out of the leaf due to water pressure inside the plant.

1 How is a root hair cell adapted to its function?

2 In what ways is osmosis similar to diffusion, and in what ways is it different?

3 Why is diffusion of water vapour out of stomata not an example of osmosis?

Roots do more than just anchor a plant in the ground – they take up water and mineral salts from the soil. The surface of roots have specialised **root hair cells** with long thin extensions that reach into the surrounding soil. The root hairs provide a large surface area for substances to enter the root.

ResultsPlus
Watch Out!

Students often confuse diffusion and osmosis. A variety of substances move by diffusion but osmosis is the movement of only water and needs a partially permeable membrane.

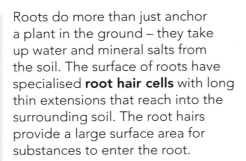

cytoplasm
nucleus
vacuole
root hairs
water
membrane of root hair cell
cell wall of root hair cell
soil particles

B A root hair cell

Osmosis and active transport

Water enters the root hair cells by **osmosis**. In this process, water moves across a **partially permeable membrane** (a membrane that only certain substances can pass through) from a region where water is in higher concentration to a region where it is in lower concentration.

In Figure C water molecules will move from solution B, where they are in higher concentration, to solution A, where they are in lower concentration. The water moves by diffusion. The sucrose molecules cannot move between the solutions because they are too big to fit through the gaps in the partially permeable membrane.

Roots can absorb nitrate and other mineral ions dissolved in soil water even if their concentration is higher in the plant than in the soil water. Absorbing particles *against* a concentration gradient is called **active transport**. The energy needed for this process comes from respiration.

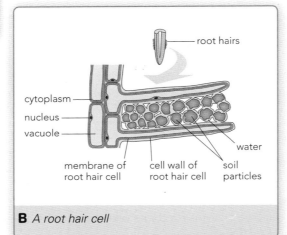

concentrated sucrose solution
dilute sucrose solution
partially permeable membrane
water molecules passes through not sucrose
sucrose molecules
water molecules
Before osmosis
After osmosis

C Diffusion of water molecules from a dilute to a concentrated sucrose solution across a partially permeable cell membrane (osmosis).

Plant transport systems

Once water and minerals have entered the root cells, they need to get to all the plant's tissues. Plants possess specialised tissue, called **xylem**, to transport water and dissolved mineral salts. Xylem tissue consists of long cells that die and form hollow tubes. These tubes also give support to the plant.

The glucose made in leaves by photosynthesis is converted to sucrose and transported to other parts of the plant by strands of living **phloem** tissue.

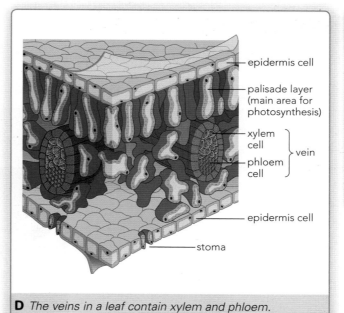

D *The veins in a leaf contain xylem and phloem.*

epidermis cell
palisade layer (main area for photosynthesis)
xylem cell
phloem cell
vein
epidermis cell
stoma

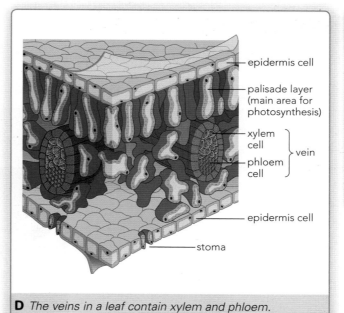

Skills spotlight

Look at the potometer in Figure E. When a fan is switched on near the plant, the air bubble moves faster. Explain how this experiment supports the idea that loss of water vapour from leaves pulls water into the plant.

Moving water against gravity

Water that evaporates from the surface of cells inside a leaf moves out of the leaf by diffusion when the stomata are open. This maintains a concentration gradient, so more water evaporates and diffuses out of the leaf. The loss of water from the leaf pulls water and dissolved mineral salts up through the xylem from the roots. This process is called **transpiration**. Factors that increase evaporation will also increase transpiration.

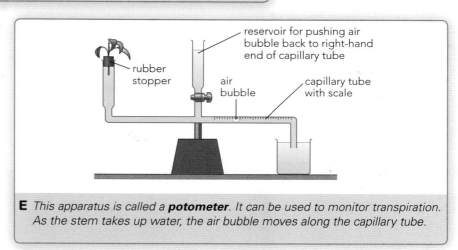

reservoir for pushing air bubble back to right-hand end of capillary tube
rubber stopper
air bubble
capillary tube with scale

E *This apparatus is called a **potometer**. It can be used to monitor transpiration. As the stem takes up water, the air bubble moves along the capillary tube.*

4 How do mineral salts get from the soil to the leaves of a plant?

5 Suggest why transpiration rates decrease at lower temperatures.

6 Explain why sunny and warm conditions usually cause the rate of transpiration to increase.

Learning Outcomes

2.17 Explain how the loss of water vapour from leaves drives transpiration

2.18 Explain how water, glucose and mineral salts are transported through a plant, including: **a** mineral uptake in roots by active transport **b** the role of xylem and phloem vessels

2.19 Describe how root hair cells are adapted to take up water by osmosis

2.20 Define osmosis as the movement of water molecules from an area of higher concentration of water to an area of lower concentration of water through a partially permeable membrane

HSW **2** Describe how data is used by scientists to provide evidence that increases our scientific understanding

>>>>>>>>>>>>>>>>>>>>> Why won't drinking seawater help shipwrecked sailors to survive?

B2.22 Investigating osmosis

How does solute concentration affect plant cells?

Plants and animals, including people, need to drink plenty of water to live, but in very rare cases too much water can kill you. A huge water overload in the body makes your blood very dilute. This causes the brain cells to swell as a result of osmosis. This can lead to seizures and even death. Too much salt water does the opposite – the salt in your blood makes water leave your body cells and they shrivel up.

A Leah Betts took an ecstasy tablet at her 18th birthday party. Ecstasy stops you feeling thirsty. Leah drank large amounts of water to try and avoid dehydration and the excess water killed her.

Water moves in and out of plant and animal cells by osmosis. Plant cells contain dissolved sugars and other chemicals. If plant cells are surrounded by water or a dilute solution of sugars, water moves into the cells by osmosis. The cells swell up and the stems are firm and support the leaves. If plant cells lose water they become floppy and shrink in size. The plant eventually wilts because the stems get thinner and the leaves cannot be held up any more.

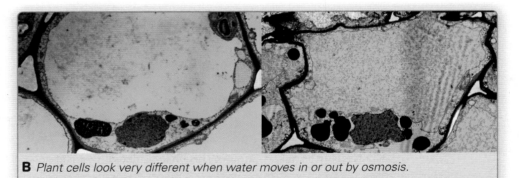

B Plant cells look very different when water moves in or out by osmosis.

Your task

You are going to plan an investigation that will allow you to find out about the effect of different strength sucrose solutions on plant cells. Your teacher will provide you with some materials to help you organise this task.

Learning Outcomes

2.21 Investigate osmosis

When planning an investigation like this, one of the skills you will be assessed on is your ability to assess the *quality of evidence*. There are 4 marks available for this skill.

Student extract 1 | A basic response for this skill

All chips were 30 mm to start with.

	Final length of potato chip (mm)		
	Sample 1	Sample 2	Sample 3
Concentrated sugar solution	25	26	24
Less concentrated sugar solution	30	25	31
Water	33	34	32

> The student has identified and dealt with an anomalous reading in the secondary evidence.

This is the table of results I found on the internet and I am using it as my secondary evidence. I think that the figure which has the ring around it does not fit with the pattern of the others and I will leave it out of any calculations I do. My own results had one odd reading where the length of a potato chip placed in water had increased by 6 mm which was more than 5% change. I think that this is also an anomalous result.

> The student has also identified where there are anomalies in their own results.

Student extract 2 | A good response for this skill

My table of results taken from the internet is given above. This has one anomalous reading. There is also one anomalous reading from my own results. I am going to use both my own results and the secondary evidence to calculate the mean change in length of my potato chips. I will leave these two anomalous results out of my calculations however because if I included them it could mean that my answer would be shifted away from the true figure. When I draw my graph I will also draw a line of best fit.

> The student has explained *why* anomalies should not be included in calculations.

> The student has explained that anomalous results will not be included in calculations.

> Drawing a line of best fit on a graph also helps to reduce the effect of anomalous results.

ResultsPlus

To access 2 marks

- Comment on the quality of your primary and secondary evidence
- Deal with any anomalies appropriately
- Say if you do not think there are any anomalies in your evidence

To access 4 marks

- Take account of any anomalies in your primary and secondary evidence
- Explain any adjustments you need to make to your evidence
- If you do not think there are anomalies, explain this and say that you are using all your evidence

Organisms and their environment

⋯ **How does the environment affect the organisms living in it?**

The sensitive plant *Mimosa pudica* has a trick to prevent its leaves being eaten. When an insect lands on it, the leaves respond rapidly by drawing water out of the cells by osmosis. The leaves fold up and the insect falls off and leaves hungry.

A *The leaves of* Mimosa pudica *fold up when touched.*

> **1** Name some factors that affect the size of a plant population. (?)

The conditions in an **environment** determine which organisms (plants, animals and microorganisms) can survive in it. The icy conditions of the Arctic and the humid heat of the rainforest each present challenges for living organisms. They must be adapted to their environments and be able to respond to any changes.

Results Plus
Watch Out!

Students find it difficult to describe biodiversity. It is the variety of species of plants and animals in an area, not the variety of individuals.

Ecologists study the **biodiversity** of life found in an **ecosystem** or **habitat**, where a particular organism is found (its **distribution**), as well as how many individuals there are in a group of the same species (the **population size**). These data are used to monitor changes in population size or distribution or to test hypotheses about what sort of organisms exist in a certain place.

Sampling means looking at a small portion of an area or population. In **random sampling** every point within an area has an equal chance of being selected. This means the sample is likely to be **representative** of the whole area.

Sampling techniques

> **2** It is important that pitfall traps are not left unchecked for too long. Why do you think this is? (?)

A **pooter** is used to catch small invertebrates through an inlet tube by sucking sharply on a second tube connected to the container. In areas with long grass a **sweep net** can be used to catch some of the organisms present. Similarly, a **pond net** can be used to sample aquatic habitats.

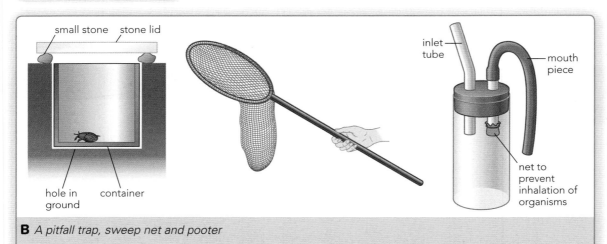

small stone   stone lid

hole in ground   container

inlet tube

mouth piece

net to prevent inhalation of organisms

B *A pitfall trap, sweep net and pooter*

Pitfall traps are useful for trapping small animals such as spiders, beetles and woodlice. They can be set up and left overnight so it is possible to catch organisms that might not be active during the day.

Quadrats are square frames of known size, which are typically used to sample the number and population of different plant species in a habitat. The quadrat is placed at random locations, for example by throwing, and the number of plants of each type that appear within the quadrat are counted. For spreading plants like clover the percentage of the quadrat area that is covered by that plant is estimated.

Random sampling is used to estimate a total population size by scaling up the mean number of one type of plant per quadrat.

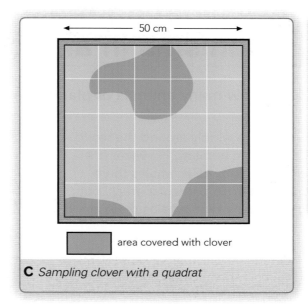

area covered with clover

C *Sampling clover with a quadrat*

3 Population density is one way to record the abundance of a organism. What do you need to know to calculate this value?

4 How would you expect light intensity to affect plant distribution?

By considering the varying conditions in a habitat, such as at different points in a field, it is possible to understand why the organisms that live there are able to do so. For example, understanding how light intensity varies might help to explain the distribution of different species of plants. Other factors such as temperature, soil or water pH may also have an effect in determining which organisms can survive in any given part of a habitat.

When investigating changes in a habitat caused by one environmental factor it is sometimes useful to carry out **systematic sampling** along a line. For example, quadrats could be placed at regular intervals along a straight line.

5 Light intensity and temperature change not only during the course of a day but also from season to season. Describe how these factors change between winter and summer and suggest how this might affect the organisms that live in a woodland habitat.

6 Suggest why scientists spend time trying to find out which organisms are present in an area before a big building project goes ahead and explain how they get a good estimate of the numbers of each species present.

Skills spotlight

It is important that scientists test their ideas. How could you set up a long-term study to test the hypothesis that use of weedkillers on a farm affects insect biodiversity of a nearby hedgerow?

Learning Outcomes

2.22 *Investigate the relationship between organisms and their environment using fieldwork techniques*

2.23 *Investigate the distribution of organisms in an ecosystem, using sampling techniques, including:*
  a *pooters*
  b *sweep nets/pond nets*
  c *pitfall traps*
  d *quadrats*
     and measure environmental factors, including:
  e *temperature*
  f *light intensity*
  g *pH*

HSW **5** *Plan to test a scientific idea, answer a scientific question, or solve a scientific problem by controlling relevant variables*

How can you count the number of whales in the ocean?

B2.24 Investigating the distribution of organisms in an ecosystem

How can you work out how many animals or plants live in a particular area?

Killer whales are distributed more widely around the world than any other animal apart from humans. However it is very difficult to count exactly how many killer whales there are. Their populations can be studied in particular areas, but scientists can only guess at how many killer whales there are in the world.

It is important to know the distribution of living organisms in an area. This is particularly true if something in the environment is going to change such as houses being built or a road developed. In these cases local environmental impact surveys are carried out. They involve collecting data about the numbers and distribution of several species in the area which will be affected.

The use of sampling techniques is an important feature of this sort of work.

Killer Whale

A The distribution of killer whales is mapped by recording every place in which a killer whale is sighted.

Your task

You are going to plan an investigation that will allow you to find out whether changes in environmental conditions affect the distribution of plants between two areas. Your teacher will provide you with some materials to help you organise this task.

Learning Outcomes

2.22 Investigate the relationship between organisms and their environment using fieldwork techniques
2.23 Investigate the distribution of organisms in an ecosystem, using sampling techniques including:
   d quadrats
   and measure environmental factors including
   e temperature f light intensity g pH

Build Better Answers

When planning an investigation like this, one of the skills you will be assessed on is your ability to *control variables.* There are 6 marks available for this skill. Here are three extracts focusing on this skill. Other skills that you need for the assessment are dealt with in other lessons.

Student extract 1 — A basic response for this skill

The student identifies a variable which must be kept constant.

> When carrying out this investigation it is important to keep the size of the quadrat the same each time. I will make sure that I use the same quadrat each time to take my results.

The student then explains how to control this variable.

Student extract 2 — A better response for this skill

The student has explained why this is an unusual investigation.

The student has not suggested how the measuring might be carried out.

> In this investigation it is hard to control variables because we are working outside the laboratory and we cannot control things like soil pH. This means that in this investigation we need to measure some of the factors so that we know what they are and can use the information when looking for patterns. For example I think that I need to measure soil pH and light intensity for each quadrat.

The student has suggested some of the factors which can be measured to produce useful results.

Student extract 3 — A good response for this skill

Gives a clear list of the factors to be measured and controlled.

Explains how these factors can be measured.

> In this investigation I need to use the same type of quadrat each time, the same system of choosing which samples to collect and the same method of counting plants. I also need to measure temperature, light intensity and soil pH. Using the same quadrat will mean that the sample size is consistent. Using a clear method to decide which areas to sample means that the sampling is random which should give us a better quality of data overall. Using the same technique for counting the plants also means that there is consistency in the results. Temperature is measured with a thermometer, pH with a pH probe and light with a light intensity meter.

Explains why it is important to be consistent with the technique.

What can fossils tell us about the history of life on Earth?

In 1912 fragments of a skull and jawbone were found at Piltdown, Sussex. Scientists thought the bones were from an ancestor of modern humans. In 1953 it was proved that the 'Piltdown Man' bones were a hoax (and were actually from modern humans and orang-utans). So ideas about human evolution had to change again.

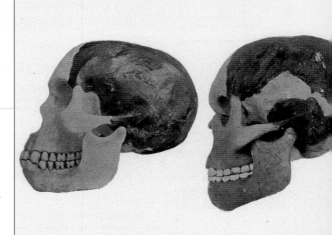

A *Different scientists put the 'Piltdown Man' bone fragments together in different ways, as they thought about what our ancestors looked like.*

1 What is a fossil?

2 Why is there no fossil record of the earliest cells?

Fossils are the preserved traces or remains of organisms that lived thousands or millions of years ago. We find fossils when the rocks containing them are weathered. The history of life on Earth as shown by the fossils from different periods of time is known as the **fossil record.** This suggests that organisms have changed gradually through time (a process called **evolution**).

The fossil record has many gaps in it. Soft tissues decay and do not usually form fossils, so soft-bodied organisms leave little fossil evidence behind. Other dead organisms did not form fossils because the hard parts were destroyed. Many fossils are buried deep in the earth and have not yet been found.

These gaps mean that scientists must interpret how organisms changed over time from incomplete data. The same sets of fossil data can be interpreted in different ways, often because fossils are frequently damaged or incomplete. The scientists who first found *Megalosaurus* fossils thought it had a huge head and walked on four legs. They hadn't found all the bones. Better fossil evidence shows it was a small-headed dinosaur walking on two legs.

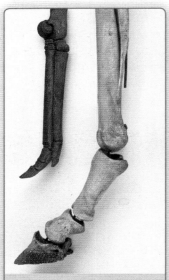

B *Differences in the limb bone fossils show how horses have evolved over time.*

3a Why are there gaps in the fossil record?
b How do these gaps affect explanations of how organisms have changed through time?

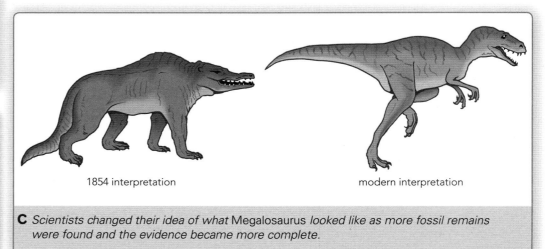

1854 interpretation   modern interpretation

C *Scientists changed their idea of what* Megalosaurus *looked like as more fossil remains were found and the evidence became more complete.*

The more fossil evidence we can collect, the better conclusions we can draw. We can date fossils accurately now and use computers to model how the organism might have looked.

More evidence for evolution

Most vertebrates have limbs, and though the limbs may look very different on the outside, the internal bone structure is very similar. This is also true of fossil vertebrates. Even fossil ancestors of limbless living vertebrates have the same basic five-fingered (**pentadactyl**) limb structure. This suggests that all vertebrates evolved from one common ancestor hundreds of millions of years ago. Looking at the fossil record, scientists can explain the evolution of different forms of pentadactyl limb in different vertebrate species as adaptations to different ways of living and moving.

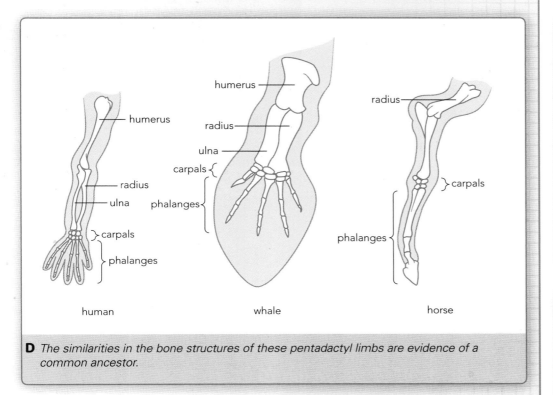

D *The similarities in the bone structures of these pentadactyl limbs are evidence of a common ancestor.*

6 Explain the problems that scientists have in interpreting the fossil record.

ResultsPlus
Watch Out!

Students sometimes give only one reason for the gaps in the fossil record. Remember that soft tissue decays so some fossils do not form and many fossils are yet to be found.

4 How is the fossil record used as evidence to support the idea of evolution?

H 5 a What are the similarities between the limbs of the different animals shown in Figure D?
b How does evolution help to explain similar features in very different organisms?

Skills spotlight

Uncertainties in scientific knowledge and ideas change over time as we find out more. Explain why modern scientists would be unlikely to be fooled by 'Piltdown Man'.

Learning Outcomes

3.1 Evaluate the evidence for evolution, based on the fossil record

3.2 Explain why there are gaps in the fossil record, including:
a because fossils do not always form
b because soft tissue decays
c because many fossils are yet to be found

H 3.3 Explain how the anatomy of the pentadactyl limb provides scientists with evidence for evolution

HSW 14 Describe how scientists share data and discuss new ideas, and how over time this process helps to reduce uncertainties and revise scientific theories

How did you grow from the size of a full stop to your current size?

How do plants and animals grow differently?

A giant redwood tree can be almost 120 m tall and over 2000 years old but it starts life as a tiny seed with a mass of only 3–5 mg.

A

ResultsPlus
Watch Out!

Describing growth as 'getting bigger' is not as good as saying that there is an increase in cell number and cell size.

Maths skills

Percentiles are used to compare a certain characteristic (e.g. mass) against the total population.

A large sample of 6-month-old babies is taken and ranked by mass. The sample is then split into 100 equally sized groups. The group with the highest masses is the 100th percentile and the group with the lowest masses is the 1st percentile. A 6-month-old baby can then be allocated a percentile depending on which group their mass falls into.

When organisms grow they get bigger. The easiest way to observe **growth** is to measure an increase in size, length or mass. For example, health professionals measure the mass and height of babies and children regularly to check that they are growing normally. Each child is compared to a chart to find out which **percentile** of the population they fall into.

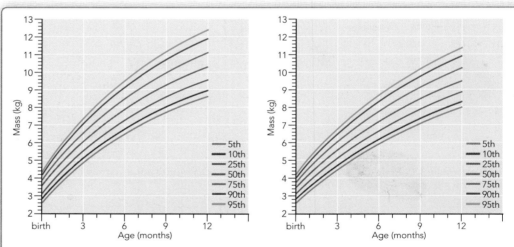

B Mass-for-age percentiles for boys (left) and girls (right), birth to 12 months. Generally babies should be between the 95th and 5th percentiles. Also if their percentile changes by two or more, this may indicate a problem.

1 A balloon gets bigger when you blow it up but this isn't growth. Why not?

2 How does cell division lead to growth?

3 How do meristems help trees grow so tall?

However, you have to be careful when you measure growth. When you blow up a balloon it gets bigger but the amount of balloon material is the same. Growth in a living organism involves two processes. There is an increase in the number of cells when a cell divides to form two identical cells, and these then get bigger.

Growth in plants

Plants grow all through their lives. Plants have special areas called 'meristems' just behind the tip of their roots and shoots where the cells keep dividing. However, growth in plants isn't just about cell division. Once the cells have divided they get longer (**elongation**).

As a plant stem or root continues to grow, the older meristem cells start to become specialised – they **differentiate.** A meristem cell can differentiate into any type of plant cell. For example, a cell in a shoot can become a leaf cell containing chlorophyll or a new cell in a root can differentiate to become a root hair cell.

Growth in animals

Growth in animals also involves cell division. But animals differ from plants because they stop growing when they become adults.

In an animal, cells that can differentiate to form a number of different types of specialised cells are called **stem cells.** These cells develop into the specialised tissues and organs of an animal's body. In an embryo, the stem cells can differentiate and form almost every type of cell needed in the body. Adults have relatively few stem cells. Each type of adult stem cell can only form a limited range of cells, such as blood cells or skeletal tissues. This is why most animals cannot re-grow a damaged limb or body part but plants can grow new shoots, roots and leaves.

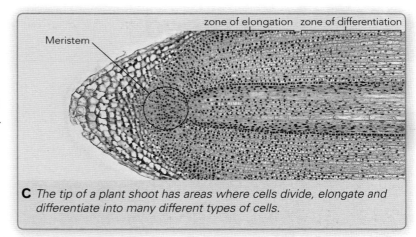

C *The tip of a plant shoot has areas where cells divide, elongate and differentiate into many different types of cells.*

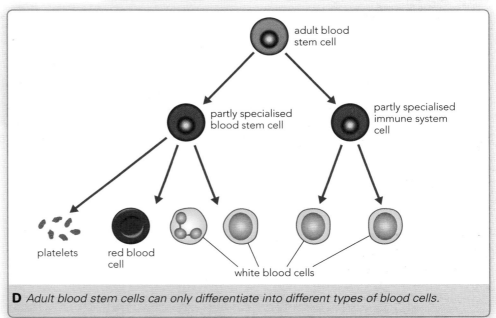

D *Adult blood stem cells can only differentiate into different types of blood cells.*

4 a List as many different cell types that you can think of which could differentiate from a human embryonic stem cell.
b How does this differ from the number of cells that can differentiate from an adult stem cell?

5 Why is it relatively easy to grow an entire plant from a single meristem cell but impossible (so far) to grow a new adult animal from a single adult stem cell?

6 Explain the similarities and differences between growth in plants and in animals.

Skills spotlight

Giant redwoods are very tall. Plan two ways of measuring the height of a giant redwood and discuss the advantages and disadvantages of each method in terms of risk and quality of data.

Learning Outcomes

3.4 Describe growth in terms of increase in size, length and mass

3.5 Interpret growth data in terms of percentile charts

3.6 Explain how cell division, elongation and differentiation contribute to the growth and development of a plant

3.7 Explain how cell division and differentiation contribute to the growth and development of an animal

HSW **5** Plan to test a scientific idea, answer a scientific question, or solve a scientific problem by choosing appropriate resources

Why is blood so important?

If you lose a lot of blood during an operation or in an accident you could die. People donate blood that can be used to help save other people's lives. A bag or unit contains 470 cm³ of blood. About 8000 units are needed every day in hospitals around the UK.

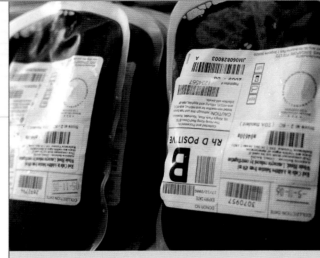

A Donated blood saves lives.

An adult has about 5 litres of blood. Although blood is a liquid, it is also an organ containing many different types of specialised cells that carry out particular functions in your body. These have all differentiated from blood stem cells and become specialised.

> **1** During what process do blood cells become specialised?

The composition of blood

Blood is made up of four main components: **plasma**, **red blood cells**, **white blood cells** and **platelets**.

Plasma is a yellow liquid. It transports dissolved substances, such as carbon dioxide, food substances and hormones.

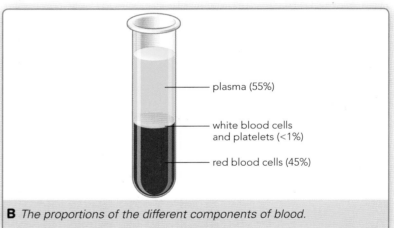

plasma (55%)

white blood cells and platelets (<1%)

red blood cells (45%)

B The proportions of the different components of blood.

ResultsPlus
Watch Out!

Students often remember the role or function of a blood cell but forget that its shape and contents help it to carry out that role. Red blood cells contain haemoglobin, which helps them to carry oxygen to the tissues.

Red blood cells contain the red pigment **haemoglobin**. Haemoglobin can combine reversibly with oxygen to form oxyhaemoglobin:

haemoglobin + oxygen ⇌ oxyhaemoglobin

When blood passes through the lungs the haemoglobin combines with oxygen. Oxyhaemoglobin is transported in red blood cells around the body to the tissues, where the oxygen is then released so that the **tissue** cells have oxygen for aerobic respiration.

> **2** Explain how red blood cells are adapted to carry as much oxygen as possible.

A red blood cell has the shape of a biconcave disc – it has a dimple on both sides (see Figure C). This adaptation gives a large surface area to volume ratio for oxygen to diffuse into and out of the cell. A red blood cell also has no nucleus and this makes room for as much haemoglobin as possible.

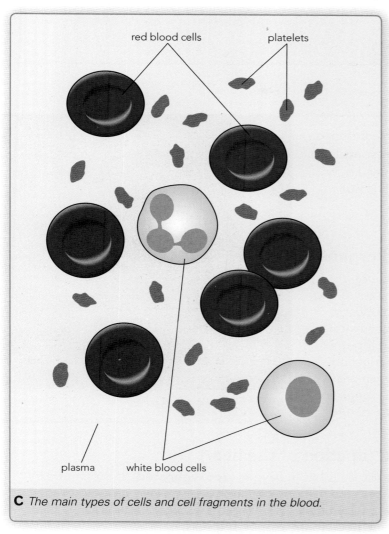

red blood cells   platelets

plasma   white blood cells

C *The main types of cells and cell fragments in the blood.*

White blood cells are part of the body's defences against disease. Some white blood cells make **antibodies**. These are proteins that bind to the microorganisms that cause disease and destroy them. Other white blood cells surround and destroy any foreign cells that get into the body. All white blood cells have a nucleus.

Platelets are tiny fragments of cells (and so do not have nuclei). They are important in making blood clot if you cut or damage your blood vessels. The clot dries out and forms a scab which also stops microorganisms getting into the body.

3 People with a disorder called thrombocytopenia do not have enough platelets. Suggest one symptom of this disease. Explain your reasoning.

4 Draw a table to show the main components of blood and their functions.

5 a Why is haemoglobin needed for cellular respiration to take place?
b Why do you think people who have just given blood may feel a bit tired?

6 Explain the following statements:
- In some diseases the red blood cells are the wrong shape and people affected will die without regular transfusions of normal blood.
- Drugs given to prevent rejection after an organ transplant can destroy the white blood cells and people affected can suffer many infections.

Skills spotlight

An argument is a discussion in which evidence is presented for and against an idea, and then used to reach a decision. In the UK only 4% of people donate blood and hospitals often get close to running out. Construct an argument against paying people for their donated blood.

Learning Outcomes

3.8 Recall the structure and function of the following parts of the blood, including:
a red blood cells
b white blood cells
c plasma
d platelets

HSW 11 Present information, develop an argument and draw a conclusion, using scientific, technical and mathematical language

Does your heart have anything to do with love?   81

How does your heart work?

We often associate the heart with our emotions. Perhaps this is because people can feel their hearts responding when they are angry or stressed. Scientists, however, know that the heart is nothing more than a pump, which is not very romantic!

A *What does a heart represent?*

> **1** Why is blood an example of an organ?

Cells, tissues and organs

A group of the same type of specialised cells forms a tissue (e.g. muscle tissue is made of muscle cells). An **organ** contains several different tissues working together to carry out a particular function in the body. The heart is an organ that pumps blood to the lungs and around the body.

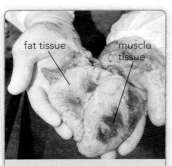

B *The heart is composed of different tissues.*

fat tissue

muscle tissue

The structure and function of the heart

Blood coming from the tissues is low in oxygen (**deoxygenated**). It is pumped by the heart to the lungs where the haemoglobin in red blood cells picks up oxygen and the blood becomes **oxygenated**. This oxygenated blood returns to the heart and is pumped around the body to the tissues and cells.

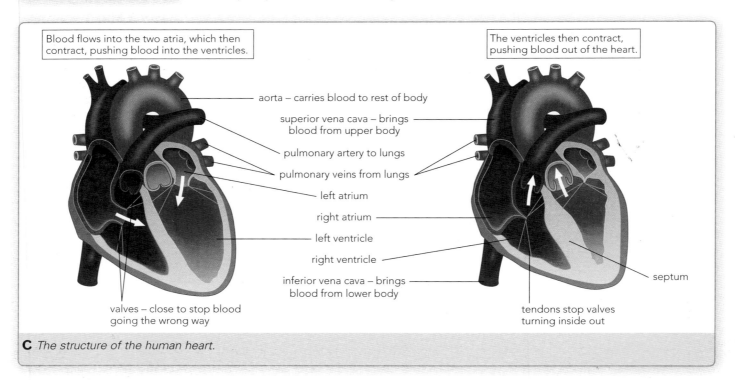

Blood flows into the two atria, which then contract, pushing blood into the ventricles.

The ventricles then contract, pushing blood out of the heart.

aorta – carries blood to rest of body

superior vena cava – brings blood from upper body

pulmonary artery to lungs

pulmonary veins from lungs

left atrium

right atrium

left ventricle

right ventricle

inferior vena cava – brings blood from lower body

septum

valves – close to stop blood going the wrong way

tendons stop valves turning inside out

C *The structure of the human heart.*

The left and right sides of the heart work together, filling and emptying at the same time. However, to explain the different parts of the heart and how they work we often follow the blood right through the heart, looking at each side in turn. The two sides of the heart are completely separated by the muscular **septum**.

A **vena cava** brings blood from the body into the **right atrium**. When the atrium is full, muscles in the wall contract and force the blood through the **valves** into the **right ventricle**. The valves are flaps of tissue that stop the blood flowing backwards.

When the ventricle is full of blood the muscles of the ventricle wall contract, forcing blood out through more valves into the **pulmonary artery**. This carries deoxygenated blood to the lungs where it picks up oxygen.

D *A dissected part of an aorta, for use in a transplant. The aorta is about the width of a garden hosepipe.*

The oxygenated blood returns from the lungs to the **left atrium** of the heart in the **pulmonary vein**. When the atrium is full, it contracts and forces the blood through valves into the **left ventricle**. Once the ventricle is full of oxygenated blood the muscles of the ventricle wall contract. This forces blood out through more valves into the **aorta**. The aorta is the large blood vessel that carries oxygenated blood around the body.

The muscle wall of the left ventricle is thicker than that of the right ventricle because it has to pump blood all around the body rather than just to the lungs.

4 a Which side of the heart pumps deoxygenated blood?
b Why is the muscular wall of the left ventricle thicker than that of the right ventricle?
c Why do you think the heart is sometimes called a double pump?

5 Describe what happens to a volume of blood from the time it goes into the heart from a vena cava until it is pumped out through the aorta.

2 What is:
a oxygenated blood
b deoxygenated blood?

ResultsPlus
Watch Out!

Students sometimes get confused when asked to describe the route of the blood through the heart. It is a good idea to start with the deoxygenated blood in the right atrium.

3 What is the function of the valves in the heart?

Skills spotlight

Ideas about how the heart works have changed over time. The second century Greek doctor Galen was not allowed to dissect human bodies. He thought that blood flowed from one side of the heart to the other through lots of invisible holes in the septum. How does this compare to modern ideas about the heart?

Learning Outcomes

3.9 Describe the grouping of cells into tissues and tissues into organs

3.10 Explain how the structure of the heart is related to its function, including:
a the four major blood vessels associated with the heart (pulmonary artery, pulmonary vein, aorta, vena cava)
b left atrium and ventricle to pump oxygenated blood
c right atrium and ventricle to pump deoxygenated blood
d valves to prevent backflow (names not required)
e left ventricle has a thicker muscle wall than the right ventricle
f the direction of blood flow through the heart

HSW 14 Describe how scientists share data and discuss new ideas, and how over time this process helps to reduce uncertainties and revise scientific theories

What is the circulatory system?

The heart pumps blood around the body all the time. If surgeons have to operate on a person's heart, they need to keep it still. The heart is stopped and a heart–lung machine takes over. It oxygenates the blood and moves it round the body while the operation is being carried out.

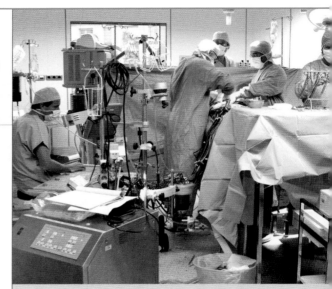

A *An artificial heart–lung machine in action.*

The heart pumps blood around the body so that all the cells in the body can get the oxygen and nutrients that they need. However, it is not the only organ that helps to do this. **Blood vessels** are also needed.

Blood vessels are tube-shaped organs that carry blood. There are three types: **arteries**, **capillaries** and **veins**.

Arteries carry blood away from the heart. The blood in arteries has to be under high pressure so that it can reach all parts of the body. So arteries have strong, thick walls.

Capillaries allow substances to diffuse into and out of the blood into, the cells in tissues. To help this process capillaries have very thin walls.

Veins have wide passages inside them and carry blood to the heart. They are wide because the blood flows relatively slowly under low pressure.

ResultsPlus
Watch Out!

Students often muddle up the directions in which arteries and veins carry blood. Remember that *Arteries* carry blood *Away* from the heart.

---

1 What is the name for all the tubes that carry blood around the body?

2 a What is an artery?
b Explain why an artery is an organ.

3 a What is the function of capillaries?
b How are capillaries adapted to this function?
c Why is this function important?

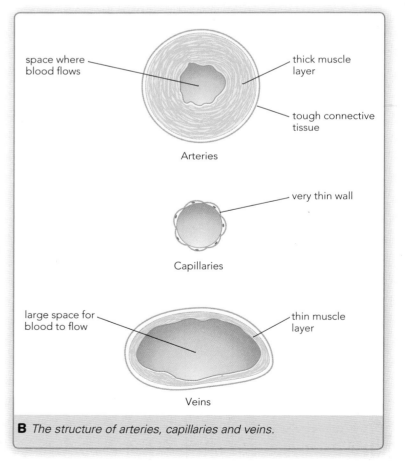

space where blood flows

thick muscle layer

tough connective tissue

Arteries

very thin wall

Capillaries

large space for blood to flow

thin muscle layer

Veins

B *The structure of arteries, capillaries and veins.*

Organ systems

Groups of specialised cells working together are called tissues and tissues that work together form organs. Groups of organs that work together are called **organ systems**. The heart and blood vessels form an organ system called the **circulatory system**.

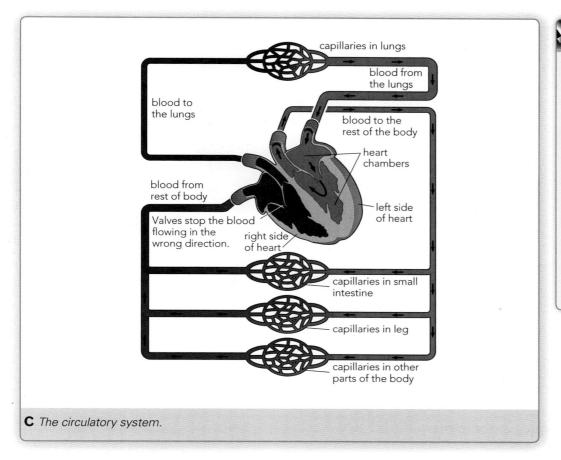

capillaries in lungs

blood from the lungs

blood to the lungs

blood to the rest of the body

heart chambers

blood from rest of body

left side of heart

Valves stop the blood flowing in the wrong direction.

right side of heart

capillaries in small intestine

capillaries in leg

capillaries in other parts of the body

C *The circulatory system.*

4 What is an organ system?

5 What is the function of the circulatory system?

6 Which type of blood vessel usually carries deoxygenated blood?

7 Describe the route of a small volume of blood from a capillary in your muscle where it has just given up its oxygen until it returns to the same place loaded with oxygen again. Include in your answer what substances pass into or out of the blood at each stage.

Learning Outcomes

3.9 Describe the grouping of cells into tissues, tissues into organs and organs into organ systems

3.11 Describe how the circulatory system transports substances around the body, including:
   a arteries transport blood away from the heart
   b veins transport blood to the heart
   c capillaries exchange materials with tissues

HSW   3 Describe how phenomena are explained using scientific theories and ideas

What is the digestive system?

Around the world snakes, ants, fresh blood mixed with milk and witchetty grubs are just some of the treats people enjoy eating. Whatever we eat, we need to be able to digest it.

A *Witchetty grubs (a type of moth caterpillar) are eaten by Aboriginal people living in the deserts of Australia.*

> **1** What happens in your digestive system?

Food contains a lot of large insoluble molecules that cannot get into the blood. These need to be broken down into small soluble molecules such as glucose, which can pass into the blood and so be used in cells.

> **2** Explain why the digestive system is an organ system.

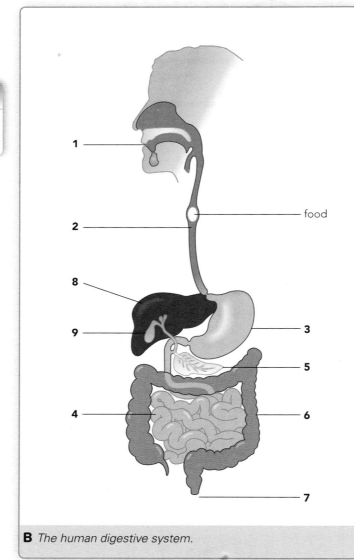

B *The human digestive system.*

Food is broken down in a process called **digestion**, which takes place in an organ system called the **digestive system**. The digestive system is made up of the **alimentary canal**, a muscular tube running through the body from mouth to anus, and several other organs that make chemicals needed for digestion (including **enzymes**).

1 Mouth: where food is taken into the body. During chewing, teeth break up the food into small pieces. This increases the surface area for digestive enzymes to work on. The tongue helps to form the chewed food into a ball called a **bolus**. The bolus gets coated in **saliva**, which lubricates it and makes it easier to swallow. Saliva also contains an enzyme that starts to break down the starch in food.

2 Oesophagus: a muscular tube between the mouth and the stomach. Muscles contract in waves to squeeze the food down towards the stomach. This is called **peristalsis**.

3 Stomach: a muscular bag that makes acid and some enzymes (most importantly, enzymes to digest proteins). It churns the food up with these juices by peristalsis to make a thick paste.

4 Small intestine: a long, coiled, muscular tube where most of the large insoluble food molecules are broken down into smaller soluble molecules. It contains lots of digestive enzymes made by the pancreas and it makes its own digestive enzymes as well. The molecules of digested food are absorbed into the blood using finger-shaped projections called **villi**, which contain capillaries. Food is moved along by peristalsis.

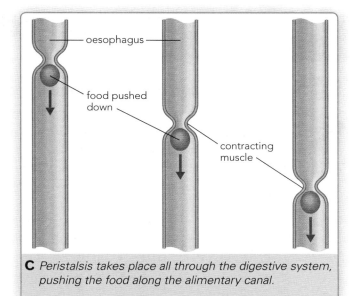

C Peristalsis takes place all through the digestive system, pushing the food along the alimentary canal.

5 Pancreas: this organ makes digestive enzymes and releases them into the first part of the small intestine.

6 Large intestine: undigested food passes into this wide, thin-walled tube. Water diffuses back into the blood leaving the waste material (**faeces**) behind.

7 Anus: where the undigested food is passed out of the body.

8 Liver: digested food is absorbed by the small intestine and dissolves in the blood plasma. Once in the blood, it is taken to the liver to be processed. Some of the molecules are broken down even more. Some are built up into larger molecules again. The liver also makes **bile**, which helps in the digestion of fats.

9 Gall bladder: a small organ that stores the bile made by the liver and releases it into the small intestine when it is needed.

> **ResultsPlus**
> **Watch Out!**
>
> Bile is *stored* in the gall bladder but it is *made* in the liver.

Skills spotlight

Some scientists measured the mass of food eaten each day by a group of 14 volunteers. They also measured the mass of faeces the volunteers produced each day. Describe how you would process and present the data produced by this investigation. You may find it helpful to include a sketch.

---

3 **a** Name the parts of the alimentary canal.
b Name the other organs that are part of the digestive system.

4 What is peristalsis and why is it so important in the digestive system?

H 5 Suggest three organs in which you would expect to find bile.

6 Compare the functions of the stomach, the small intestine and the large intestine.

Learning Outcomes

3.12 Describe the functions of the parts of the digestive system, including: **a** mouth, **b** oesophagus, **c** stomach, **d** small and large intestines, **e** pancreas, **f** liver, **H g** gall bladder

3.13 Explain the role of the muscular wall of the alimentary canal in peristalsis

HSW *11* Present information using scientific conventions and symbols

>>>>>>>>>>>>>>>>>>>>>>>>>>>> Why could the contents of your stomach damage your skin?

How is your food digested by enzymes?

Skimmed milk has very little fat, while Jersey cream has lots of fat suspended in it. The difference shows in the colour and taste. Milk and cream are both examples of emulsions – lots of tiny fat particles suspended in a watery liquid.

A

jersey cream   skimmed milk

> **1** What substances bring about the chemical breakdown of food?

The digestive system breaks down large insoluble food molecules into smaller, soluble molecules. This chemical breakdown of food depends on digestive enzymes. Different types of digestive enzyme break down the three main types of food molecules: **carbohydrates**, **proteins** and **fats**.

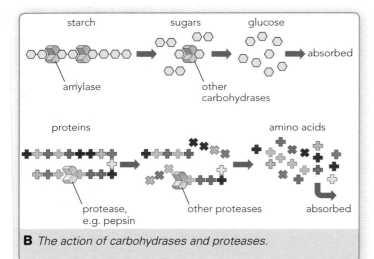

starch   sugars   glucose

absorbed

amylase   other carbohydrases

proteins   amino acids

protease, e.g. pepsin   other proteases   absorbed

B *The action of carbohydrases and proteases.*

Digesting carbohydrates

Foods like bread and potatoes are full of carbohydrates (substances made of carbon, hydrogen and oxygen). The simplest carbohydrates are called **sugars** and these can be built up into more complex carbohydrates, such as **starch**. Digestive enzymes that break down carbohydrates are known as **carbohydrases**. **Amylase** is a carbohydrase that breaks starch down into sugars, which can then be absorbed by the small intestine or broken down into glucose by other carbohydrases.

An amylase is present in saliva. Another amylase is made in the pancreas and released into the small intestine.

Digesting proteins

> **2 a** What carbohydrase breaks down starch?
> **b** What are the products of this breakdown?

Proteases are enzymes that digest proteins, breaking them down into shorter chains and then into **amino acids**. **Pepsin** is a protease made in the stomach. It works best in acidic conditions. The stomach walls produce an acid, which makes the pH 2–3. This is the optimum pH for pepsin to break down protein as fast as possible. However, the contents of the small intestine are alkaline and so the proteases released into the small intestine work best at about pH 8.

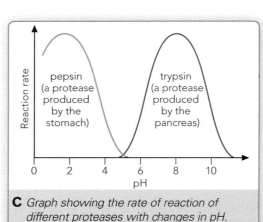

Reaction rate

pepsin (a protease produced by the stomach)

trypsin (a protease produced by the pancreas)

0   2   4   6   8   10
pH

C *Graph showing the rate of reaction of different proteases with changes in pH.*

Digesting fats

Lipases are enzymes that digest fats. Lipases chemically break down fat molecules into **fatty acids** and **glycerol**.

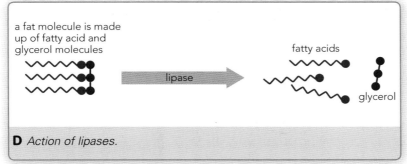

a fat molecule is made up of fatty acid and glycerol molecules

fatty acids

lipase

glycerol

D *Action of lipases.*

Fat and water don't mix so the fats and oils you eat form globules in the watery digestive juices. Large globules have a very small surface area to volume ratio, which means the lipases can only break down the fat molecules very slowly. However, bile physically breaks down the large globules into tiny droplets, forming an **emulsion**. We say that bile **emulsifies** the fat. The smaller droplets have a larger surface area, which makes it possible for lipase to break down the fat molecules far more rapidly. More bile is released after a fatty meal. Bile makes faeces brown.

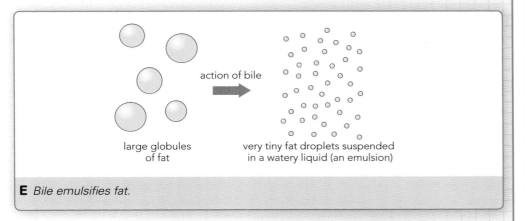

large globules of fat

action of bile

very tiny fat droplets suspended in a watery liquid (an emulsion)

E *Bile emulsifies fat.*

The bile from the gall bladder is alkaline. This helps to neutralise the acid from the stomach and produces a slightly alkaline environment for the protease enzymes of the small intestine to work in.

H 5 Describe the two roles of bile in the digestive system.

6 Explain the role of enzymes in digesting food. Include examples of different enzymes in your answer.

3 a What is a protease?
b Where does pepsin work?
c What is the optimum pH for pepsin and how does the stomach provide this environment?
d Why would the proteases found in the small intestine not work well in the stomach?

4 Why is an alkaline environment important in the small intestine?

Skills spotlight

A simple model of how large molecules are broken down during digestion uses a scissors analogy to imagine the enzyme chopping the molecule into smaller ones. How well does this model explain what happens?

ResultsPlus
Watch Out!

If you remember that fats are called lipids it might help you to remember that lipase is the enzyme that breaks them down. A word ending in *ase* is usually an enzyme.

Learning Outcomes

3.14 Explain the role of the digestive enzymes, including:
  a carbohydrases, including amylase, which digest starch to simple sugars
  b proteases, including pepsin, which digest proteins to amino acids
  c lipase which digests fats to fatty acids and glycerol

H *3.15* Explain the role of bile in neutralising stomach acid and emulsifying fats

HSW *3* Describe how phenomena are explained using scientific models

>>>>>>>>>>>>>>> How could you fit something with the same area as a tennis court inside your body?

How do digested food molecules get into the blood?

These breads and cakes look delicious but the food on the right would make people with coeliac disease very ill. Their bodies react to gluten, a protein found in cereals like wheat and barley, and this damages the lining of the small intestine. This is why gluten warnings are important on food labels.

A *Some people can't eat food containing gluten. The food on the left is gluten free but the food on the right contains wheat.*

1 a We say that digested food is absorbed. What does 'absorbed' mean?
b Where does absorption occur?

Large insoluble food molecules are broken down into small soluble molecules by digestive enzymes. These products of digestion must be absorbed from the small intestine into the blood, to be delivered to the cells where they are needed.

small intestine

villi

Large molecules cannot be absorbed.

network of capillaries

Digested molecules can be absorbed.

The covering of the villus is only one cell thick.

blood supply

B *The structure of a villus.*

Absorbing food in the small intestine

Digested food passes into the blood by diffusion through the intestine and capillary walls. The bigger the surface area available, the more diffusion can take place.

The small intestine is a tube about 5 m long and 2.5 cm in diameter. If the inside surface was smooth the surface area would only be about 0.5 m² and very little absorption would take place. However, the lining of the small intestine has millions of finger-like folds called **villi** that make the surface area much bigger. If it were opened out, the surface area would actually be about 200 m² (tennis-court sized!). This means much more diffusion can take place.

Each villus has a good network of blood capillaries. This means there is always a lower concentration of soluble food molecules in the blood than there is inside the small intestine because soluble food molecules are constantly moved away in the blood. This steep **concentration gradient** between the two areas means diffusion takes place rapidly down it. Finally, there is only a single layer of cells between the contents of the small intestine and the blood vessels in the villi. This means there is only a short distance over which the soluble food molecules need to diffuse. The large surface area, good blood supply and short distances all make the diffusion of dissolved food molecules into the blood as efficient as possible.

Evidence for the importance of villi

Evidence that the villi are important for increasing the efficiency with which the soluble products of digestion are absorbed into the blood can come from studying coeliac disease. In this disease the villi may be lost. People affected cannot absorb the products of digestion properly. They often become very thin as a result.

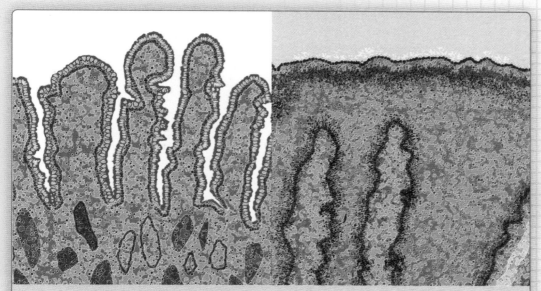

C *The effects of coeliac disease on villi in the small intestine can clearly be seen in the picture on the right.*

4 a Look at Figure C. What is the difference between the normal lining of the gut and the gut affected by coeliac disease?
b Use your answer to part **a** to suggest how coeliac disease can be used as evidence for the importance of a large surface area for the absorption of digested food.

5 Explain how the presence of villi in the gut makes the absorption of the soluble products of digestion so efficient.

2 What adaptations does the small intestine have for absorption?

3 a Explain what a concentration gradient is.
b What effect does it have?

Skills spotlight

Some students make a model gut out of a membrane that lets soluble food molecules through. They put starch solution and amylase inside the 'gut,' and show that sugars move out of the gut by diffusion into the surrounding water. Discuss some of the uses and limitations of this simple model of a gut.

ResultsPlus
Watch Out!

When asked about absorption in the small intestine, students often say that it has a large surface area. They forget to mention that it is very long and has lots of villi and that each villus has a large surface area.

H **3.16** Explain how the structure of villi (large surface area, single layer of cells and capillary network) allows efficient absorption of the soluble products of digestion

 **3** Describe how phenomena are explained using scientific models

B2.33 Enzyme concentration

Does the rate of starch digestion depend on the concentration of amylase?

You produce about 1.5 litres of saliva every day. The concentration of carbohydrase enzymes in the saliva varies from person to person.

Saliva contains the enzyme **amylase**, a carbohydrase which starts to break down starch into simple sugars. Starch digestion starts in the mouth and is completed in the small intestine. Once the starch is fully digested, the simple sugars diffuse into your blood through the villi. The simple sugars are carried around your body to be used in the cells.

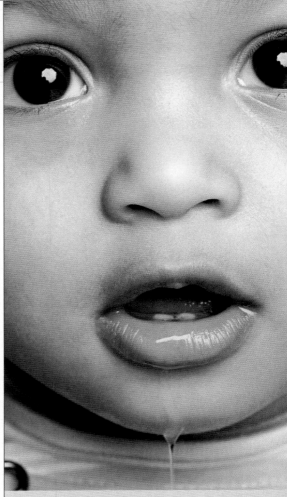

A *Babies often dribble saliva, but as we get older we keep it in our mouths.*

Your task

You are going to plan an investigation that will allow you to find out the effect of different concentrations of an enzyme on the breakdown of food. You will need to use a model of the alimentary canal and evaluate how well it represents what really happens. Your teacher will provide you with some materials to help you organise this task.

Learning Outcomes

3.17 Investigate the effect of different concentrations of digestive enzymes, using and evaluating models of the alimentary canal

When planning an investigation like this, one of the skills you will be assessed on is your ability to draw *conclusions based on evidence*. There are 6 marks available for this skill. Here are two extracts focusing on this skill. Other skills that you need for the assessment are dealt with in other lessons.

Student extract 1 | **A typical response for this skill**

There is an important point missing here – the student should have said if this conclusion supports their hypothesis or not.

> I think that increasing the concentration of the enzyme speeds up the rate of reaction, but only up to a certain point. I think that this is a proportional relationship because when one factor (the concentration) increases the other (the time) decreases by the same amount each time. I can see this from my graph which shows a steady fall in time taken for a positive result, up to a certain point and then it slows down and flattens off. I can also see this from the graph I found in the text book which shows how the rate of reaction changes with the concentration of the enzyme for another enzyme controlled reaction.

The student has talked about the relationship between the two variables.

This is good use of both primary and secondary information.

Student extract 2 | **A good response for this skill**

This is good because it refers to the way that the conclusion supports the hypothesis.

To access higher marks you need to use your scientific knowledge to explain what you think is happening in the investigation.

> I think that increasing the concentration of the enzyme speeds up the rate of reaction, but only up to a certain point. I can see this from my graph which shows a steady fall in the time taken for a positive result up to a certain point and then flattens off. This supports my hypothesis that changing the concentration of the enzyme would change the rate of reaction. This is a directly proportional relationship for the first part of the graph. This also fits with my scientific knowledge that enzymes are biological catalysts which speed up reactions. When there are more enzymes there are more catalysts to speed up the rate of the reaction. The rise in the rate of reaction will stop when there are enough enzyme molecules for each molecule of substrate. Adding more enzyme after this point will have no effect on the rate of the reaction – it is going as fast as possible. I can also see this from the graph I found in the text book which shows how the rate of reaction changes with the concentration of the enzyme for another enzyme controlled reaction.

 ResultsPlus

To access 4 marks

- Provide a conclusion based on all your collected evidence
- Explain your conclusion
- Describe any relevant mathematical relationships in your conclusion

To access 6 marks

You also need to:
- Refer to the original hypothesis in your conclusion
- Refer to other scientific ideas in your conclusion

Probiotics and prebiotics

Do probiotics and prebiotics improve your health?

People affected by cystic fibrosis make thick sticky mucus that can stop their pancreas releasing enzymes into the small intestine. They have to take replacement enzymes to help them digest their food.

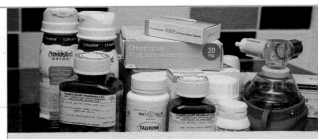

A *The enzymes that someone with cystic fibrosis must take every day.*

Many healthy people now eat **functional foods** as well as their normal food. These foods claim to make you healthier.

Probiotics

The digestive system contains about 10^{13} bacteria – that's more bacteria than you have cells in your body! Some of these bacteria can cause problems but most provide health benefits. They help break down food and protect against disease-causing microorganisms.

1 Suggest one reason why people might eat functional foods. ?

Probiotics contain live bacteria often called 'friendly' or 'beneficial' bacteria. These are usually *Lactobacillus* and *Bifidobacteria*, which produce lactic acid in the gut. The manufacturers of probiotic foods such as yogurts or yogurt drinks claim that these will make you healthier by improving your digestive system, helping your body protect itself against disease and reducing allergies. However, in 2010 scientists at the European Food Safety Agency looked at the evidence for 180 health claims for probiotics. They rejected 10 completely and said there was not enough evidence to support the other 170 claims.

2 What do probiotic foods contain? ?

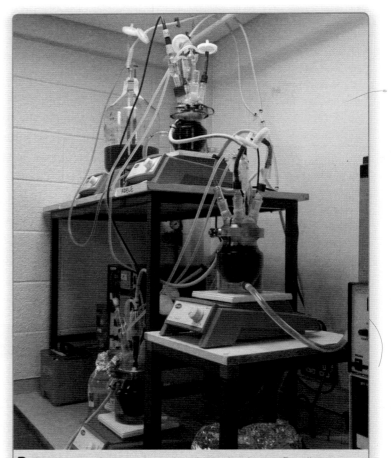

B *The model of a large intestine that scientists at Reading University use to investigate the effects of functional foods.*

Plant stanol esters

Plant stanol esters are oily substances found in plants. Scientists have discovered that these can stop the small intestine absorbing **cholesterol**, lowering the levels of cholesterol in the blood. High cholesterol levels are linked to a raised risk of heart disease. Plant stanols are now used in many foods such as yogurt, drinks and spreads. There is clear evidence that they have an effect.

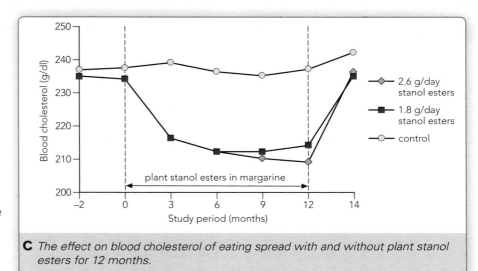

C *The effect on blood cholesterol of eating spread with and without plant stanol esters for 12 months.*

Prebiotics

Prebiotics are substances that the body can't digest. They act as food for the 'beneficial' bacteria in the gut and encourage their growth. Tomatoes, bananas, onions and asparagus all contain **oligosaccharides**, a common form of prebiotic. You also find prebiotics in specially made dairy products or sold in capsules. The evidence is growing that prebiotics can increase the beneficial bacteria in your gut and so help maintain good health. The graph in Figure D shows the effect of adding an oligosaccharide to the diet on the recurrence of diarrhoea in a group of 142 patients.

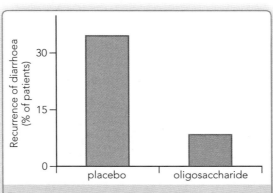

D *The effect on diarrhoea of adding an oligosaccharide to the diet, from a study of 142 patients*

3 a What do prebiotic foods contain?
b How are prebiotics claimed to work?

4 What is a plant stanol ester?

5 How would you evaluate the data given in Figure D on the impact of oligosaccharides on health?

6 a Explain what is meant by a functional food.
b Looking at Figure C, what do you think the control group ate?
c Explain what other controls would need to be in place in the diet.
d Is it worth taking the higher dose of plant stanols? Explain your answer.

7 A new functional food for toddlers is being advertised on TV, claiming to increase the numbers of 'good bacteria' in the gut. How would you evaluate these claims?

Skills spotlight

People need to be able to rely on the science used in adverts to help them make decisions. In 2009 an advert for a probiotic yogurt drink claiming that it was 'scientifically proven to help support your kids' defences' was banned. Why do you think it was banned?

Learning Outcomes

3.18 Evaluate the evidence for the claimed benefits of the use of functional foods as part of a healthy diet, including:
   a probiotics containing *Bifidobacteria* and lactic acid bacteria *Lactobacillus*
   b prebiotic oligosaccharides
   c plant stanol esters

HSW **13** Explain how and why decisions about uses of science and technology are made

These questions are indicative of the type of questions used in the exam. Refer to page 6 for information on the grades.

Plants in the environment

1. The diagram shows a root hair cell.

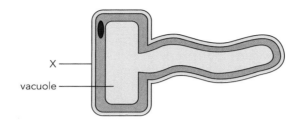

(a) (i) Part X is the

A cell membrane
B cytoplasm
C cell wall
D chloroplast

(1)

(ii) Describe the role of the vacuole in the root hair cell. (2)

(iii) The root hair cells take in water from the soil. The plant uses this water for photosynthesis.

Describe how water moves from the soil into the root hair cell. (2)

(b) (i) Susan was investigating the number of nettle plants growing in a 10 m² area of overgrown land next to her school. She counted 2 nettle plants in a 0.5 m² quadrat, which she placed randomly on the land. How many nettle plants in total did Susan estimate there were in the 10 m² area of land? Show your working. (2)

(ii) Susan compared the size of the nettle leaves growing in the overgrown land to those growing in a nearby wood.

nettle leaf found in the wood

nettle leaf found in overgrown land

Explain the differences in the size and colour of the two leaves. (3)

Exercise and heart rate

2. Beverlyn carried out an investigation to test how different types of exercise affect pulse rate. The graph shows Beverlyn's pulse rate before each exercise and for 1 minute after each exercise.

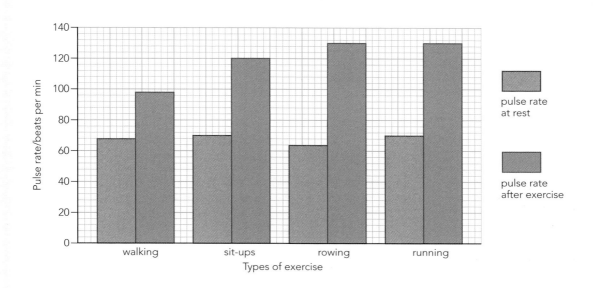

(a) (i)   Which exercise made Beverlyn's pulse rate increase the most?   (1)

(ii) By how much more does Beverlyn's pulse rate increase with sit-ups compared to walking? Show your working.   (2)

(b) (i)   Describe how Beverlyn's body is able to get enough oxygen to her working muscles during exercise.   (3)

(ii) Beverlyn's muscles use oxygen to help provide her body with the energy she needs to continue exercising. Which process takes place inside Beverlyn's muscles that uses oxygen to release energy?

   A anaerobic respiration
   B aerobic respiration
   C gas exchange
   D excess post-exercise oxygen consumption   (1)

(c) Explain why Beverlyn's heart rate remains high for a short time after she has stopped exercising.   (3)

DNA

3. DNA is a double helix with each strand linked together by a series of paired bases. There are four bases in a DNA molecule.

   (a) The table shows the percentage of each base found in a sample of DNA from a kiwi fruit.
   Copy and complete the table to give the names of the two missing bases.

Percentage of bases (%)	Name of base
24.4	thymine
25.6	cytosine
24.4	
25.6	

(1)

   (b) (i) A section of a DNA molecule contains 1000 bases. 40% of the bases are thymine. How many cytosine bases are contained in this section of a DNA molecule? Show your working. (2)

   (ii) The bases in a DNA molecule are linked together by

   A strong hydrogen bonds
   B weak hydrogen bonds
   C strong covalent bonds
   D weak covalent bonds (1)

   (c) Rosalind Franklin was a scientist who completed work in the 1950s on DNA structure. At the same time Francis Watson and James Crick were working on similar ideas. When Watson and Crick heard about Franklin's work they thought they may have made a mistake. Suggest two ways in which scientists could find out if there was a mistake in their work. (2)

   (d) Many people believe that Watson and Crick would not have completed their work if Franklin had not been doing similar work. Suggest two ways in which her work may have helped them. (2)

Photosynthesis

4. The diagrams show the results of an experiment to find out how light intensity affects photosynthesis.

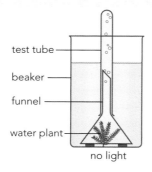
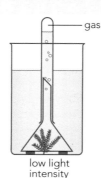
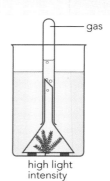

test tube
beaker
funnel
water plant

no light

low light intensity

high light intensity

gas

gas

(a) (i) Use the words from the box to complete the following sentences.

| carbon dioxide   oxygen   nitrogen |
| low light intensity   high light intensity   no light |

The gas produced during this experiment was _____.

During this experiment the most gas was produced at _____. (2)

(ii) During photosynthesis the reactants required are

    A glucose and oxygen
    B glucose and water
    C carbon dioxide and water
    D carbon dioxide and oxygen (1)

(b) An experiment was carried out on the factors that affect the rate of photosynthesis in a well-watered plant. The graph shows the results of the experiment. Throughout the experiment the temperature was kept at the optimum level for enzymes in the plant.

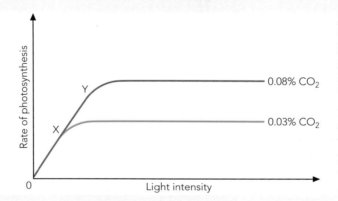

Explain why, at the higher CO_2 concentration, light is the limiting factor on photosynthesis between point X and point Y. (3)

(c) This is a diagram of the structure of a leaf. Describe how this leaf is adapted for photosynthesis. (6)

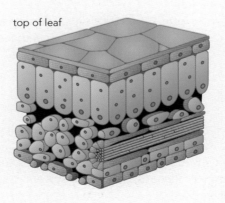

top of leaf

The digestive system

5. The diagram shows the human digestive system.

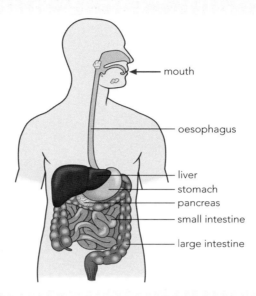

mouth

oesophagus

liver
stomach
pancreas
small intestine

large intestine

(a) (i) Draw **one** line from each part of the digestive system to its correct function. One has already been done for you.

Part of digestive system

Function

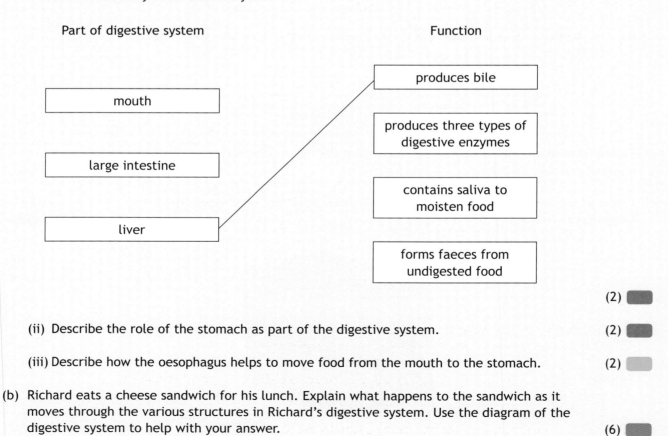

mouth

large intestine

liver

produces bile

produces three types of digestive enzymes

contains saliva to moisten food

forms faeces from undigested food

(2)

(ii) Describe the role of the stomach as part of the digestive system. (2)

(iii) Describe how the oesophagus helps to move food from the mouth to the stomach. (2)

(b) Richard eats a cheese sandwich for his lunch. Explain what happens to the sandwich as it moves through the various structures in Richard's digestive system. Use the diagram of the digestive system to help with your answer. (6)

Cloning the woolly mammoth

1. Scientists have unearthed body cells of a woolly mammoth that had been preserved in ice for thousands of years. These cells contain frozen DNA samples that could be used to clone the woolly mammoth.

(a) (i) The woolly mammoth has been extinct for over 13 000 years. One problem with cloning a woolly mammoth is that

   A there are no live sperm cells to fertilise the female egg cells
   B the frozen DNA contains the haploid number of chromosomes
   C the body cells found contain the diploid number of chromosomes
   D there is no suitable surrogate to implant a re-nucleated egg   (1)

   (ii) The diagram shows how a parent cell containing the DNA of a woolly mammoth divides to produce two new cells.

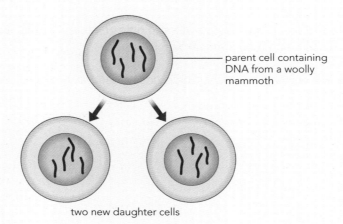

parent cell containing DNA from a woolly mammoth

two new daughter cells

   Explain how the parent cell containing DNA from a woolly mammoth divides to produce the two new daughter cells shown in the diagram.   (3)

   (iii) Describe how the two new daughter cells continue to divide and grow to produce a woolly mammoth.   (2)

   (iv) Describe the stages involved in cloning a woolly mammoth.   (3)

The role of enzymes

2. The diagram illustrates how a certain enzyme functions.

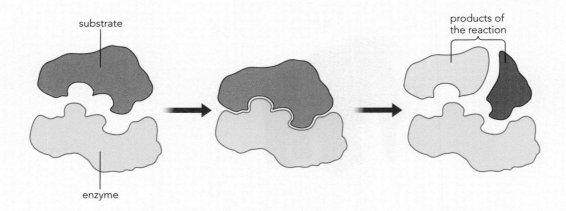

substrate

products of the reaction

enzyme

(a) (i) Enzymes are described as biological catalysts. Describe what the term biological catalyst means. (2)

(ii) What is the name of the area on the enzyme where the substrate binds?

    A active site
    B enzyme-substrate complex
    C enzyme-product complex
    D double helix (1)

(b) (i) Enzymes are put in biological washing powders to help to clean clothes. Explain how specific enzymes act to clean food stains from clothes. (3)

(ii) Explain why biological washing powders should not be used at temperatures above 40°C. (3)

The mammalian heart

3. The diagram shows a mammalian heart.

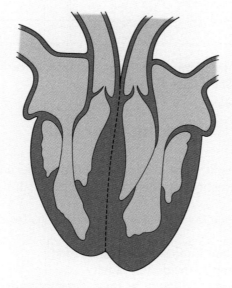

(a) (i) The blood vessel that brings deoxygenated blood back from the body into the heart
is called the

    A vena cava
    B pulmonary vein
    C pulmonary artery
    D aorta (1)

  (ii) Copy the diagram and draw arrows to show the direction of blood flow through
both sides of the heart. (1)

(b) Describe how the action of the heart ensures that blood is oxygenated before it is
delivered to body cells. (2)

(c) 'Plant stanol esters can help to keep the heart healthy'. Describe the scientific
claim behind this statement. (2)

(d) The cardiac output (CO) is the amount of blood, in litres, ejected by the left side of the heart
every minute. CO is calculated by multiplying the stroke volume (SV, the amount of blood in
millilitres ejected per heartbeat) by the heart rate (HR) in beats per minute.

$$CO = SV \times HR$$

The table shows information on the SV, HR and CO of an adult at rest and of the same adult performing strenuous exercise. The SV for strenuous exercise has not been recorded.

	SV ml per beat	HR beats per min	CO l per min
at rest	70	70	4.9
strenuous exercise		154	20

What is the difference between the stroke volume of the adult at rest and the adult performing strenuous exercise? Show your working. (3)

Making new cells

4. Adult stem cells are important for maintaining and repairing the tissues and organs that make up a living organism. Scientists are currently investigating whether adult stem cells can be used for transplants.

   (a) Which of the following is a disadvantage of using adult stem cells for transplants?

   A   they are unable to differentiate into the cell type needed
   B   they have no Hayflick limit and so divide too rapidly
   C   obtaining large numbers of cells is difficult as their capacity to divide is limited
   D   embryos are destroyed once research is complete (1)

   (b) The DNA found in the nuclei of stem cells contains the information for making the cell's proteins. These are necessary for cell differentiation, growth and division and also for the production of gametes.

   Describe how the process of producing gametes is different from the process that produces body cells for growth. (3)

   (c) Haemoglobin is a protein found in red blood cells. It contains 2 alpha polypeptide chains. The DNA sequence for each alpha chain contains 330 base pairs. What is the total number of amino acids found in both alpha chains of haemoglobin? Show your working. (2)

   (d) Describe the roles of mRNA and tRNA in protein synthesis. (6)

The circulatory system

5. Blood is a mixture of plasma, white blood cells, red blood cells and platelets. The diagram shows a microscope slide of a blood smear.

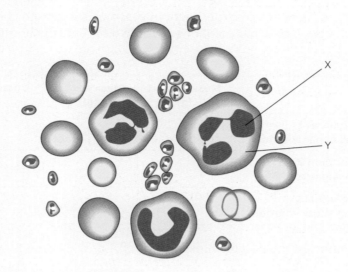

(a) (i) How many white blood cells are shown in the diagram? (1)

(ii) What part of the cell is indicated by label **X**?

    A haemoglobin
    B cytoplasm
    C nucleus
    D cell membrane (1)

(iii) What part of the cell is indicated by label Y?

    A haemoglobin
    B cytoplasm
    C nucleus
    D cell membrane (1)

(b) Describe how the structure of a red blood cell is related to its function. (3)

(c) Describe the role of the heart in the transport of oxygen into and around the body. (6)

Here are three student answers to the following question. Read the answers together with the comments around and after them.

Joanna's teacher asked her to carry out an experiment into osmosis in potato chips. She placed potato chips of the same size and mass into sugar solutions of three different concentrations. Her results are shown below.

	Mass at start (g)	Mass at end (g)	Change in mass (g)
Solution 1 (very weak sugar solution)	40	48	8
Solution 2 (medium sugar solution)	40	41	1
Solution 3 (very strong sugar solution)	40	34	−6

Explain how osmosis has effected the potato chips in solution 1 and solution 3.

Student answer 1 | Extract typical of a level ① answer

The question asks you to explain: this just repeats information from the table.

This is wrong: think about what you know about osmosis to help you work out the answer.

In solution 1 the potato chip gets bigger by 8 and in solution 3 the chip gets smaller by -6. Solution 2 doesnt show much change. I thing the chip gets bigger because the sugar goes into the chip.

Remember to stick to the question – there are no marks for talking about solution 2.

Summary
This is a level 1 answer because it only describes the information. The description is mostly clear and correct, but the question asks for an explanation. The examiner wants you to give scientific reasons for these results. An explanation would need to include a definition of osmosis as this was introduced in the question.

Student answer 2 — Extract typical of a level ② answer

A good start but you need to explain in much more detail why the water moved out. You should not just describe the data.

> Osmosis is the movement of water from where it is high to where it is low through a membrane. In solution 1 the chip gained weight because the water went in to the chip. In chip 3 the chip lost weight because water went out. The sugar didn't move.

This is quite a good definition of osmosis and a good way to start the answer. However more detail is needed, especially relating to concentration gradients.

Could be a useful detail if it explained why the sugar didn't move.

Summary
This answer suggests that you have a fair grasp of the topic. It starts to give an explanation but falls short of being scientific and explaining why the water is moving relative to the concentration gradient. The answer needs to give more detail about what is happening in the experiment.

Student answer 3 — Extract typical of a level ③ answer

Good explanation but it would be clearer if it referred to differences in concentration rather than stronger and weaker.

> Osmosis is the movement of water molecules from an area of higher water concentration to an area of lower water concentration across a partially permeable membrane. In solution 1 the chip gets heavier because water moves into the chip where the sugar is stronger. The chip becomes turgid. In solution 3 water comes out of the chip because the sugar solution is stronger outside. The chip becomes plasmolysed.

A good explanation of osmosis.

Good use of scientific terms here.

Summary
A good clear answer to the question with detail and use of scientific terms to explain what is happening in the experiment. The answer outlines the definition of osmosis and shows that the student has the scientific knowledge as well as the ability to apply that knowledge.

 ResultsPlus

Move from level ① to level ②

To move from level 1 to level 2 you must first use the information you have been given in the table, but also give some explanation of why this is happening. A good idea when the question introduces a scientific process like osmosis is to try and define the process using as much detail as you can. Try to use scientific terms as much as possible but do not use terms incorrectly, or you may be penalised for wrong science.

Move from level ② to level ③

The level of scientific detail is crucial here. Be careful to use scientific terms and explain the results as clearly as you can with reference to the subject matter. A clear definition of the process of osmosis is vital in gaining the level 3. You must have a coherent flow to the answer and you must ensure that you include as much information as you can to show the examiner your breadth of knowledge. Terms such as concentration gradient, turgid and plasmolysed, if used correctly, will raise your level significantly.

Here are three student answers to the following question. Read the answers together with the comments around and after them.

| Question | Microscopy and cell structure | Grade | G–C |

Microscopy has helped scientists to understand the structure and function of cells in much more detail.

Describe the functions of the various components of plant and animal cells and explain how the use of microscopes can help scientists to understand their structures in more detail.

Student answer 1 — Extract typical of a level ① answer

A simple explanation has been given for the use of microscopes.

> Microscopes make cells bigger so scientists can see them. Plants have got cloropasts for photosinthis and cloraphyll. Animals have got skin cells and they are bigger than plants and they don't have cell walls.

The answer correctly states that plant cells contain chloroplasts and chlorophyll for photosynthesis.

Although the answer says that animal cells do not have a cell wall, it is not true that animal cells are generally larger than plant cells.

Summary
This answer gives very simple, brief information, particularly on how to use the microscope. It could be improved by adding more detail on the structure and function of cells. Some of the scientific terms have been spelt wrongly – you will gain marks for spelling these correctly.

Student answer 2 — Extract typical of a level ② answer

The answer includes a similarity and a difference between plant and animal cells and in both cases states the function of the components mentioned.

> Plant and animal cells are similar because they both have a nucleas that controls the cell. Plants have cloroplasts for photosynthesis to make their food, but animals don't. When you shine the light through cells it magnerfies them so they look larger. You can change the lens to 40 times to make the cell look bigger. Animals have skeletons but plants have cell walls which support them and they both have cell membranes. Both kinds have ribasomes and mitachondria.

Good – scientific terminology is used when explaining how the light microscope can be used to study cells.

Another correct comparison between plant and animal cells with a correct function for the cell wall.

Summary
Some of the information in this answer is good, and it covers both the structure and function of cells. However, the information is not well organised. The description of how a microscope works does not explain the role of lenses. The answer could be improved by including more aspects of microscopy than just magnification, e.g. resolving power and the use of electron miscroscopes. Take care with terminology – some of it has been spelt incorrectly.

This covers a good range of cell components and describes their functions clearly.

Plant and animal cells have similar components – the cell membrane lets certain substances enter and exit the cell whereas the nucleus controls the cell activities. Only plants have a cell wall which provides structural support and chloroplasts containing chlorophyll which traps light energy for photosynthesis. In plant cells, the vacuole contains water which also supports the structure of the cell. Both types of cell have mitochondria which is where energy is released for use in chemical reactions. Cells can be magnified using microscopes such as the light microscope although to see lots of detail microscopes such as the SEM are used. Different objective lenses can be used to magnify the cells by different amounts without distorting the image so they have a high resolving power. The objective lenses in a light microscope allows the image to be magnified up to about 100 times but at this magnification the image is not really sharp.

Good – this includes two different types of microscope that can be used to help in the study of cells.

There is quite a bit of detail on how the image can be magnified and the benefits of electron microscopy over light microscopy are explained well, by discussing resolving power.

Summary

This is a clear level 3 response that discusses the function of a range of cell components as well as how the microscope can be used to help in the study of cells. It is clearly structured and scientific terminology is used correctly throughout.

 ResultsPlus

Move from level ❶ to level ❷

To move to level 2 you should describe the functions of a greater range of cell components. You should also give a fuller answer to the first part of the question, using scientific terminology to explain how a microscope makes images larger.

Make sure that your answer is spread equally between the two parts of the question and make sure that your spelling is accurate.

Move from level ❷ to level ❸

To move to level 3, you need to organise the answer well, with one section on comparing animal and plant cells, and another on the use of microscopes. Make sure to describe the functions of cell components rather than just list them. The information on microscopy should explain in detail how different types of microscopes work to help scientists study cells.

Here are three student answers to the following question. Read the answers together with the comments around and after them.

Question — Mitosis and meiosis — Grade E–A*

At fertilisation the gametes join together to form a zygote. Explain the roles of the two types of cell division, mitosis and meiosis, in the formation of gametes and the subsequent cell division following fertilisation.

Student answer 1 — Extract typical of a level ① answer

It is a good point to recognise that sperm and ova are gametes.

It would be better to say that gametes have half the number of chromosomes as body cells.

> Mitosis is wat happens in animals growing and repairing. Sperm is a gamete and so is the egg. These are made by miosis. They have less chromosones.

The explanation of meiosis is correct, but try not to mis-spell scientific terms, especially when they are given in the question.

Summary

This is a very basic answer. The gametes are correctly identified and so are the different types of cell division. However, there is no explanation of the processes involved during the cell division or the differences between the two types.

Student answer 2 — Extract typical of a level ② answer

No details are given about what happens in meiosis. If you leave out the detail you will lose marks.

> Mitosis is when cells divide when tissues are growing or repairing. Meiosis is for the production of gametes – sperm and ovum. During mitosis the DNA lines up in the middle of the cell and makes copies of itself. The DNA moves to the sides of the cell and the cell divides. In meiosis the same thing happens but it don't copy itself. At the end of meiosis there are four cells.

Some stages of mitosis are explained but some are missing. Try to include as much detail as possible.

Summary

For level 2 one of the types of cell division should be described in detail or there should be some detail about both. In this case there is some detail about both types of cell division, but in places it is vague. Mitosis is described fairly well, although some stages are missing. Meiosis is identified as gamete formation but the actual mechanism of meiosis is not described. There is no mention of the fact that the gametes are haploid (have half the normal number of chromosomes).

This is a little jumbled but overall it makes sense and the detail of the chromosomes in the cell is accurate.

Mitosis is used for growth and tissue repair. Cell division by mitosis results in 2 new cells identical to the original cell. Chromosomes copy themselves before cell division, then they line up in the middle and are pulled to the side and the cell divides. •——

This is a good description of mitosis.

Good level of detail.

In meiosis the gametes are formed and they have half the number of chromosomes of the original cell. The first bit of the process is the same as mitosis and then the cells divide again making four cells and each of them is different. •——

Summary

This is a good answer and overall the content is accurate. More detail could have been included, such as whether the cells are haploid or diploid. There could also have been more detail of the actual cell division. However, it is a clear and concise answer that meets the level 3 requirements.

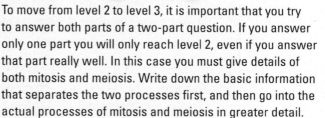

Move from level ① to level ②

To move from level 1 to level 2 you need to include detail of some of the stages in mitosis or meiosis. If you are unsure of both, give as much detail as you can on one of them. Make sure you keep to the facts and only put down the information you are sure about. Including wrong information could mean you stay at level 1.

Move from level ② to level ③

To move from level 2 to level 3, it is important that you try to answer both parts of a two-part question. If you answer only one part you will only reach level 2, even if you answer that part really well. In this case you must give details of both mitosis and meiosis. Write down the basic information that separates the two processes first, and then go into the actual processes of mitosis and meiosis in greater detail. That way, if you run out of time you could still include enough information to get to level 3.

Here are three student answers to the following question. Read the answers together with the comments around and after them.

Question | Transpiration | Grade | E–A*

The graph shows the amount of water taken up by a plant and the amount of water lost by the plant through transpiration in one day.

Explain the patterns in water loss by transpiration and water uptake over one day.

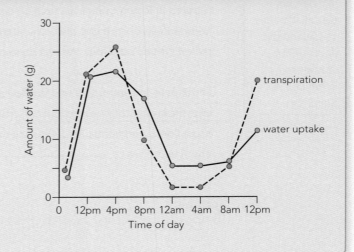

Student answer 1 | Extract typical of a level ① answer

The question asks for an *explanation* of the patterns in the graph. Simply stating that there is a pattern will not gain any marks.

The answer correctly identifies a similar pattern in the two graphs, but it doesn't explain why this pattern is seen.

the graphs has the same pattern. They are both high at 12 noon on both days and low between 12 and 4 at night. There isn't as much water uptake at night as there is in the day although theres quiet a lot at 8 at night time compared to 12 at night. This is probaly because its colder at night and the plant dosent do photosyntheses at night and water is needed for photosyntheses.

The student seems to be saying that the amount of transpiration and water uptake is related to temperature and levels of photosynthesis, but this is not explained clearly.

Summary

The response gives very simple, brief information but lacks scientific detail to explain the patterns shown by the graph. Although there is some indication that the amount of water taken up and lost is related to temperature and the rate of photosynthesis, this is not explained very clearly or in much detail. Spelling and grammar is generally poor.

Student answer 2 | Extract typical of a level (2) answer

Good: this links water uptake with osmosis and identifies that transpiration is a process in which plants lose water.

There is no explanation here of why less water is lost at night.

The water uptake increases up to about 26g by 12 oclock on the first day but transpiration does not increase as much so the plant takes in more water by osmosis than it looses. The plant carries out photosynthesis during the day so it needs more water although as it gets dark it dosent need as much water so the amout it takes in goes down. Transpiration also goes down when it gets darker. It goes down more between midnight and 4 oclock which means that it isn't loosing as much water at night compared to day time.

More water is taken up than is lost through transpiration: this is an important point.

Another good point: the plant needs water during the day for photosynthesis. Water uptake is therefore lower at night.

Summary

This answer gives some explanation of water uptake and loss by the plant. Some scientific information is included to support the information given, and some scientific terminology is used to explain the patterns in the graph. You explain that the rate of photosynthesis influences water uptake and loss, but important factors such as temperature and humidity are not mentioned. There are some inaccuracies in spelling and grammar.

Student answer 3 | Extract typical of a level (3) answer

This links water uptake with photosynthesis, and also makes the point that high temperatures increase transpiration, so the plant needs even more water.

This is the first answer to describe how a decrease in temperature reduces transpiration.

The amount of water lost through transpiration is similar to water uptake by osmosis from the soil. During the hottest part of the day, between about midday and 4 pm, water uptake peaks. Plants use water for photosynthesis but they also lose it from their leaves so they need to take in more water. The amount of water lost during this time is also greatest, although it begins to level off slightly. This is could be because the stomata start to close to stop the plant losing too much water. When evening comes, water uptake goes down a lot, and so does the amount of water lost through transpiration. At midnight both are at their lowest. This is because there is no photosynthesis to use water. The temperature is lower as well, which slows down evaporation of water from the leaves.

This is a good possible explanation of why the amount of water lost through transpiration might level out during the hottest part of the day.

Summary

This is an excellent answer that clearly explains water uptake and water loss through transpiration. Scientific terminology is used well throughout and it is written in a coherent and logical fashion. There is little detail missing and the answer shows a good understanding of the topic.

Results Plus

The diagram below shows how bases on a length of DNA relate to the amino acids in the final protein produced. Use the diagram to help you explain how DNA controls the making of specific proteins in a cell. (3)

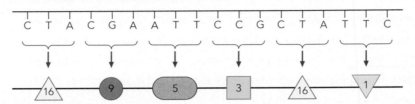

Student's answer: A triplet codes for an amino acid.

This would get one mark. Only about half of students obtained any marks at all.
There are three marks available for this question, so you need to make at least three points in your answer. The different colours in this answer show each separate marking point. If more marks were allowed, this answer would have gained seven marks! If you had written at least three of these sections, in the correct order and without making any incorrect statements, you would have gained the three marks. Less than a fifth of students got full marks.

DNA unzips and the coding strand is used to make mRNA. The mRNA moves to the ribosomes where tRNA is formed. The tRNA fetches amino acids, coded for by a triplet on the tRNA, and the protein is assembled from a string of amino acids.

Results Plus
Exam question report

The strands of a DNA molecule are linked by pairs of bases. Which of these is a correct pair?

A adenine with guanine   **B** guanine with thymine   **C** thymine with cytosine   **D** cytosine with guanine

Answer: The correct answer is D.

How students answered

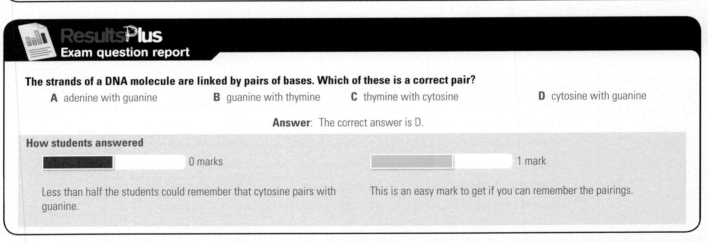

0 marks

Less than half the students could remember that cytosine pairs with guanine.

1 mark

This is an easy mark to get if you can remember the pairings.

Results Plus
Exam question report

Trees need minerals such as nitrates to grow effectively. How do trees obtain these minerals?

A by osmosis using respiration   **B** by osmosis using photosynthesis   **C** by active transport using respiration   **D** by active transport using photosynthesis

Answer: The correct answer is C.

How students answered

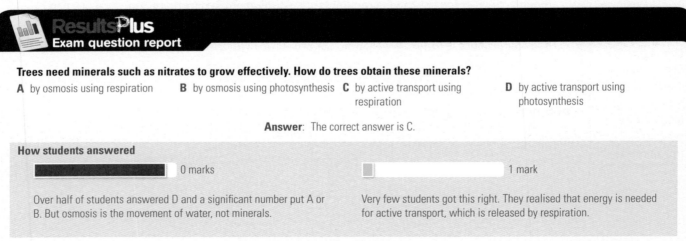

0 marks

Over half of students answered D and a significant number put A or B. But osmosis is the movement of water, not minerals.

1 mark

Very few students got this right. They realised that energy is needed for active transport, which is released by respiration.

A tennis player's breathing rate increases during a game. Explain why the breathing rate increases. (3)

 As there are three marks, you need to give three points. A good answer would be:
To get more oxygen to the muscles for more aerobic respiration, as they are working harder.
It is important to say *more* oxygen is needed for *more* respiration. An answer that just refers to needing oxygen, or needing lots of oxygen, would not get the marks. We need lots of oxygen even when we are not exercising!
(You could also mention that it is to remove more carbon dioxide to prevent a build-up of lactic acid or reduce the oxygen debt (EPOC).)
Many students get confused about what happens to heart rates and breathing rates during exercise. For this question, only 6% of students gained all three marks.

A farmer thinks he can increase the yield of tomatoes by increasing the light intensity in his greenhouses. An advisor suggests to the farmer that another limiting factor could be preventing further increases in photosynthesis.
Name one limiting factor and describe how its effect could be reduced. (2)

Student's answer: carbon dioxide.

Examiner's comment: This answer only gets one mark because it does not describe how the effect of the limiting factor could be reduced.

A good answer would be: 'Carbon dioxide. Increase the amount of carbon dioxide by putting paraffin heaters in the greenhouse.'

Newts are small animals that live in ponds. A newt can regenerate a leg if one is accidentally cut off. The diagrams show the regeneration of the leg over a 24-day period.

leg cut off here

stump

(a) Name the type of cell division that results in the regeneration of a leg. (1)

Correct answer: mitosis. Most students got this right.

(b) Describe the changes that occur in the stump to allow the leg to regenerate. (2)

Correct answer: Any two points from the following: A scab forms to seal the wound. Cells at the stump become undifferentiated stem cells. These cells divide. The dividing cells differentiate to form the different kinds of cells needed in the new leg.

Wrong answers included:
- a detailed description of mitosis (these students hadn't read the question carefully enough, and assumed that part b was asking about the same thing as part a
- descriptions of the diagram (such as 'The leg grows and the toes are made'), which weren't detailed enough to gain any marks.

State one way that mRNA is different from DNA. (1)

Student's answer: RNA is not all of DNA.

Examiner's comment: This answer is not specific enough to get the mark.

A good answer would be: RNA has a single strand but DNA has a double strand.
Other acceptable answers would be: DNA stays in the nucleus but some RNA leaves the nucleus, or RNA has the base uracil instead of thymine.

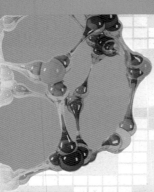

Chemistry 2
Discovering chemistry

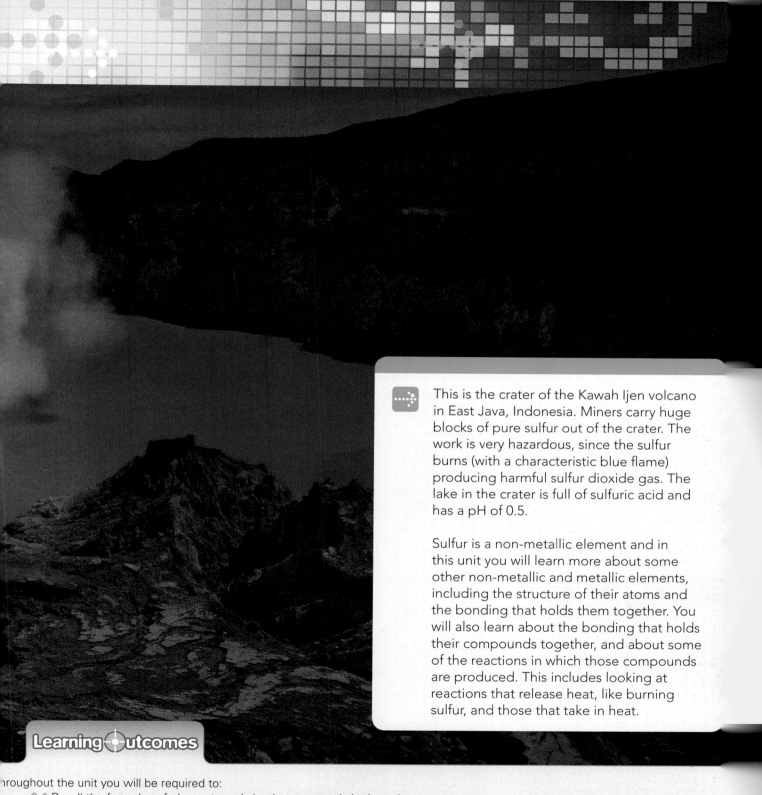

This is the crater of the Kawah Ijen volcano in East Java, Indonesia. Miners carry huge blocks of pure sulfur out of the crater. The work is very hazardous, since the sulfur burns (with a characteristic blue flame) producing harmful sulfur dioxide gas. The lake in the crater is full of sulfuric acid and has a pH of 0.5.

Sulfur is a non-metallic element and in this unit you will learn more about some other non-metallic and metallic elements, including the structure of their atoms and the bonding that holds them together. You will also learn about the bonding that holds their compounds together, and about some of the reactions in which those compounds are produced. This includes looking at reactions that release heat, like burning sulfur, and those that take in heat.

Learning Outcomes

hroughout the unit you will be required to:

0.1 Recall the formulae of elements and simple compounds in the unit

0.2 Represent chemical reactions by word equations and simple balanced equations

0.3 Write balanced chemical equations including the use of state symbols (s), (l), (g) and (aq) for a wide range of reactions in this u

0.4 Assess practical work for risks and suggest suitable precautions for a range of practical scenarios for reactions in this unit

0.5 Demonstrate an understanding that hazard symbols used on containers:
 a indicate the dangers associated with the contents
 b inform people about safe-working procedures with these substances in the laboratory

How did Dmitri Mendeleev make his periodic table of elements?

By the 1860s 63 elements were known. Chemists were keen to make a table that organised the elements in a helpful way, but this proved difficult. In one table of elements, every eighth element had similar properties. Unfortunately, this worked only as far as calcium so it was not a success.

Then in 1869, the Russian chemist Dmitri Mendeleev (1834–1907) was busy writing the second volume of his chemistry textbook. He was having trouble deciding which elements to write about next. Mendeleev's solution was to construct a table.

A A monument to Mendeleev in St Petersburg.

Mendeleev's table is now called the **periodic table**. To make his original table, Mendeleev used the latest measurements of atomic masses (called 'weights' then) available. He also carefully considered the properties of the different elements.

1 What information about the elements did Mendeleev use to develop his table?

Mendeleev arranged the elements in order of increasing atomic mass. Other chemists had tried this before, but Mendeleev sometimes broke this rule. He did this so that elements with similar properties lined up in his table. For example, iodine should come before tellurium according to its atomic mass. Mendeleev swapped the positions of these two elements so that they lined up with elements with similar properties.

Leaving gaps

Mendeleev put elements with similar properties into horizontal rows in his first table. However, he continued to work on his table. By 1871, he settled on a table in which elements with similar properties were organised into vertical columns, just as in the modern periodic table. Unlike other chemists, Mendeleev thought that there must be elements still to discover, so he left gaps for them.

ResultsPlus
Watch Out!

Remember that in a periodic table the periods are the rows and the groups are the columns.

Series	Group 1	Group 2	Group 3	Group 4	Group 5	Group 6	Group 7	Group 8
1	H1							
2	Li 7	Be 9	B 11	C 12	N 14	O 16	F 19	
3	Na 23	Mg 24	Al 27	Si 28	P 31	S 32	Cl 35.5	
4	K 39	Ca 40	? 44	Ti 48	V 51	Cr 52	Mn 55	Fe 56 Ce 59 Ni 59 Cu 63.5
5	(Cu 63.5)	Zn 65	? 68	? 72	As 75	Se 79	Br 80	
6	Rb 85	Sr 88	Y 89	Zr 91	Nb 93	Mo 96	? 100	Ru 101 Rh 103 Pd 106 Ag 108
7	(Ag 108)	Cd 112	In 115	Sn 119	Sb 122	Te 128	I 127	

2 What, in Mendeleev's table, suggests that he was not sure where to put copper (Cu) and silver (Ag)?

B Mendeleev's 1871 table with modern chemical symbols and atomic masses. The red boxes are gaps left for elements not known at the time.

Making predictions

Mendeleev used the gaps in his table to make predictions about the properties of undiscovered elements. One of these predictions was for an element that he called eka-aluminium. When gallium was discovered shortly afterwards in 1875, its properties closely fitted those he had predicted for eka-aluminium.

Property	Eka-aluminium (Ea)	Gallium (Ga)
atomic mass	About 68	70
density of solid (g/cm³)	6.0	5.9
melting point	Low	29.8 °C
formula of oxide	Ea_2O_3	Ga_2O_3
density of oxide (g/cm³)	5.5	5.88
reacts with acids and alkalis?	Yes	Yes

C *The predicted properties of eka-aluminium and the actual properties of gallium*

D *The main part of the modern periodic table, highlighted to show the locations of **metals** (yellow) and **non-metals** (blue).*

7 Describe two ways in which Mendeleev's arrangement of elements was new at the time. For each feature, explain what he did and why it was useful. Include relevant examples.

3 Explain why the discovery of gallium was seen as a successful test of Mendeleev's periodic table.

4 a What was the general order in which Mendeleev arranged the elements in his table?
b Explain why Mendeleev put tellurium before iodine in his table, instead of the other way round.

5 Use the periodic table to find out if krypton, Kr, is a metal or non-metal.

Skills spotlight

Scientists sometimes have ideas (or inspiration) that allow them to make new steps in scientific understanding. Identify two new ideas Mendeleev had to develop his table and how the evidence supported these ideas.

6 Suggest why Mendeleev predicted that eka-aluminium would be a metal and not a non-metal.

Learning Outcomes

1.1 Explain how Mendeleev: **a** arranged the elements, known at that time, in a periodic table by using properties of these elements and their compounds **b** used his table to predict the existence and properties of some elements not then discovered

1.2 Classify elements as metals or non-metals according to their position in the periodic table

HSW 2 Describe the importance of creative thought in the development of hypotheses and theories

What is inside an atom?

Atoms are incredibly tiny. It is not possible to see an individual atom, let alone see what is inside one. Figure A is not a photo but a computer image made by a microscope that measures the electrical resistance between an incredibly fine electrode and a surface. The information is used by a computer to make an image of the surface.

1 Assuming that a nucleus is 20 000 times smaller than an atom, calculate the diameter of the nucleus of an atom which is scaled up to 100 m in diameter.

Atoms are the smallest particles of an element that can take part in chemical reactions. They are too small to see directly. However from the results of many experiments to understand their structure, scientists have concluded that each atom contains an even smaller object at its centre, called the **nucleus**. This is about 20 000 times smaller than the atom itself.

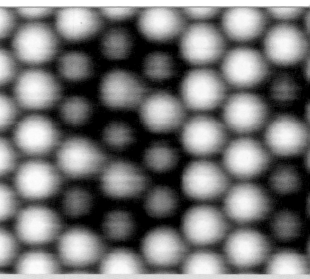

A The chemical symbol for silicon 'written' by individual silicon atoms surrounded by tin atoms. The colours wer added by the computer to make the image clearer.

B Most of an atom is empty space. The diameter of the London Eye is 135 m. If an atom could be made the same size, its nucleus would be about the size of a pea.

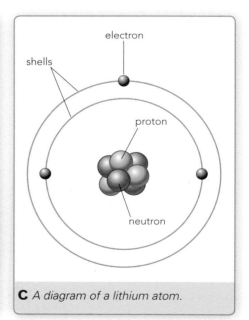

C A diagram of a lithium atom.

2 What can you say about the size of the nucleus shown in Figure C compared to the size of the lithium atom?

Subatomic particles

Atoms are made from three types of subatomic particle. The nucleus contains **protons** and **neutrons**. It is surrounded by **electrons**. The electrons are arranged in **shells** (or **energy levels**) at different distances from the nucleus.

3 How many protons, neutrons and electrons are there in the lithium atom shown in Figure C?

Subatomic particles have very, very small masses and electrical charges. Rather than give their actual masses and charges, it is more convenient to describe their masses and charges compared to a proton. These are called the particle's **relative mass** and **relative charge**.

Subatomic particle	Relative mass	Relative charge
proton	1	+1
neutron	1	0
electron	negligible ($\frac{1}{1840}$)	−1

D *The masses and charges of subatomic particles, compared to the mass and charge of a proton.*

Atoms and elements

Every atom of a particular element has the same number of protons. No two elements have the same number of protons in their atoms. For example, all hydrogen atoms contain one proton, and all helium atoms contain two protons. All atoms contain the same number of protons as electrons. This means that they have no overall charge.

6 An aluminium atom has 13 electrons and 14 neutrons.
a How many protons does it have?
b Explain why an aluminium atom is electrically neutral.
c An atom has 14 protons and 14 neutrons. Explain why it cannot be an aluminium atom.

7 Imagine you have to describe the structure of an atom to a friend on the phone without the use of picture or video messaging. Write down what you would say.

4 Which subatomic particle has the lowest mass?

5 Where is most of the mass of an atom found?

Skills spotlight

In an experiment in 1909, positively charged particles were fired at gold atoms. Most of them went straight through the atoms, but some were repelled at an angle. This was evidence for a positively charged nucleus at the centre of the atom. Suggest why most of the particles went straight through the gold atoms.

Learning Outcomes

1.3 Describe the structure of an atom as a nucleus containing protons and neutrons, surrounded by electrons in shells (energy levels)

1.4 Demonstrate an understanding that the nucleus of an atom is very small compared to the overall size of the atom

1.5 Describe atoms of a given element as having the same number of protons in the nucleus, where this number is unique to that element

1.6 Recall the relative charge and relative mass of: a a proton b a neutron c an electron

1.7 Demonstrate an understanding that atoms contain equal numbers of protons and electrons

HSW 2 Describe how data is used by scientists to provide evidence that increases our scientific understanding

How are the elements arranged in the modern periodic table?

Mendeleev's periodic table gradually became accepted by scientists. With a few adjustments as new discoveries were made, it led directly to the modern periodic table. This is now so familiar that you see it on t-shirts and mugs. You can even sit down for a meal at one (see Figure A).

A

1 What are the atomic number and mass number of an element?

For many years, an element's atomic number was just its position on the periodic table. Then, in 1913, scientists worked out that an element's **atomic number** was the number of protons in its atoms.

The total number of protons and neutrons in the nucleus of an atom is called its **mass number**. In full chemical symbols like $^{27}_{13}Al$ the top number is the mass number and the bottom one is the atomic number.

B

mass number
(number of protons
+
number of neutrons)

$^{27}_{13}Al$

atomic number
(number of protons)

This information can be used to calculate the number of protons, neutrons and electrons in an atom. For example, in $^{27}_{13}Al$, there are:

- 13 protons (because the atomic number is 13)
- 27 − 13 = 14 neutrons (mass number minus atomic number)
- 13 electrons (the same as the number of protons).

2 Work out the number of protons, neutrons and electrons in an atom of $^{23}_{11}Na$.

Relative atomic mass

The mass of an atom is incredibly small, so we use **relative atomic mass** instead of its actual mass in kilograms. The symbol for relative atomic mass is A_r. Carbon is used as the standard atom. A carbon-12 atom contains 6 protons, 6 neutrons and 6 electrons. Its relative atomic mass is defined as exactly 12. The masses of all other atoms are compared to this. For example, the mass of a helium-4 atom is one-third that of a carbon-12 atom. So its relative atomic mass is 4.

Maths skills

To find the relative atomic mass of chlorine you need to calculate the average (**mean**) mass of its atoms, where

$$mean = \frac{sum\ of\ values}{number\ of\ values}$$

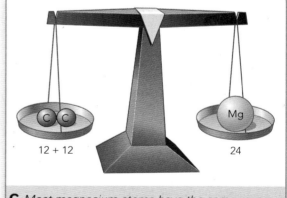

C Most magnesium atoms have the same mass as two carbon-12 atoms, so their relative atomic mass is (12 + 12) = 24.

The modern periodic table

3 How are the elements arranged in the modern periodic table?

4 Describe what groups and periods in the periodic table are.

The elements in the modern periodic table are arranged in order of increasing atomic number rather than in order of atomic mass as Mendeleev did. The horizontal rows are called **periods**; the vertical columns are called **groups**. Each group contains elements with similar properties. The main groups are numbered 1 to 7 from left to right. The group on the far right is group 0.

Group 1 2   3 4 5 6 7 0

Period																	

relative atomic mass —— 1 H
atomic number —— 1

1																	4 He 2
2	7 Li 3	9 Be 4										11 B 5	12 C 6	14 N 7	16 O 8	19 F 9	20 Ne 10
3	23 Na 11	24 Mg 12										27 Al 13	28 Si 14	31 P 15	32 S 16	35.5 Cl 17	40 Ar 18
4	39 K 19	40 Ca 20	45 Sc 21	48 Ti 22	51 V 23	52 Cr 24	55 Mn 25	56 Fe 26	59 Co 27	59 Ni 28	63.5 Cu 29	65 Zn 30	70 Ga 31	73 Ge 32	75 As 33	79 Se 34	80 Br 35	84 Kr 36
5	85 Rb 37	88 Sr 38	89 Y 39	91 Zr 40	93 Nb 41	96 Mo 42	(98) Tc 43	101 Ru 44	103 Rh 45	106 Pd 46	108 Ag 47	112 Cd 48	115 In 49	119 Sn 50	122 Sb 51	128 Te 52	127 I 53	131 Xe 54
6	133 Cs 55	137 Ba 56	139 La 57	178 Hf 72	181 Ta 73	184 W 74	186 Re 75	190 Os 76	192 Ir 77	195 Pt 78	197 Au 79	201 Hg 80	204 Tl 81	207 Pb 82	209 Bi 83	(209) Po 84	(210) At 85	(222) Rn 86
7	(223) Fr 87	(226) Ra 88	(227) Ac 89	(261) Rf 104	(262) Db 105	(266) Sg 106	(264) Bh 107	(277) Hs 108	(268) Mt 109	(271) Ds 110	(272) Rg 111							

lanthanides

140 Ce 58	141 Pr 59	144 Nd 60	145 Pm 61	150 Sm 62	152 Eu 63	157 Gd 64	159 Tb 65	162 Dy 66	165 Ho 67	167 Er 68	169 Tm 69	173 Yb 70	175 Lu 71

actinides

232 Th 90	231 Pa 91	238 U 92	237 Np 93	244 Pu 94	243 Am 95	247 Cm 96	247 Bk 97	251 Cf 98	252 Es 99	257 Fm 100	258 Md 101	259 No 102	262 Lr 103

D *The modern periodic table shows the element's symbol, relative atomic mass and atomic number.*

Skills spotlight

Scientific knowledge and ideas change over time as new evidence becomes available. Explain how the discovery that the atomic number was the number of protons in an atom led to a change in the periodic table.

ResultsPlus
Watch Out!

In exams, many students get the mass number and atomic number mixed up. Remember that the top number is the *mass* number and is always *more* than the atomic number.

Isotopes

Isotopes are different atoms of an element with the same number of protons and electrons but different numbers of neutrons. For example, chlorine atoms always have 17 protons, but some can have 18 neutrons and others have 20 neutrons. These are different isotopes – $^{35}_{17}$Cl and $^{37}_{17}$Cl.

The presence of isotopes means that some relative atomic masses are not whole numbers. For example, the **relative abundance** (proportion) of $^{35}_{17}$Cl is 75% of all chlorine atoms, and the relative abundance of $^{37}_{17}$Cl is 25%. This is why the relative atomic mass of chlorine is 35.5 rather than a whole number:

$$\text{relative atomic mass} = \frac{(75 \times 35) + (25 \times 37)}{(75 + 25)}$$

$$= \frac{2625 + 925}{100} = \frac{3550}{100} = 35.5$$

H **5 a** What are isotopes? **b** There are two isotopes of copper: 69% is $^{63}_{29}$Cu and 31% is $^{65}_{29}$Cu. Calculate the relative atomic mass of copper. Give your answer to one decimal place.

6 The relative atomic mass of titanium is 48. Explain as fully as you can what this means.

Learning Outcomes

1.2 Classify elements as metals and non-metals according to their position in the periodic table

1.8 Explain the meaning of the terms: **a** atomic number **b** mass number **c** relative atomic mass

1.9 Describe the arrangement of elements in the periodic table such that: **a** elements are arranged in order of increasing atomic number, arranged in rows called periods **b** elements with similar properties are placed in the same vertical column, called groups

H *1.10* Demonstrate an understanding that the existence of isotopes results in some relative atomic masses not being whole numbers

H *1.11* Calculate the relative atomic mass of an element from the relative masses and abundances of its isotopes

HSW *14* Describe how scientists share data and discuss new ideas, and how over time this process helps to reduce uncertainties and revise scientific theories

How are the electrons arranged in atoms?

The system of chemical symbols used today was first developed in 1819 by Jöns Jacob Berzelius (1779–1848), a Swedish chemist. Before then several different systems existed. This made it difficult for scientists to understand each other's work. It is important that scientists develop and use systems to communicate their ideas clearly, including for showing the arrangement of electrons in atoms.

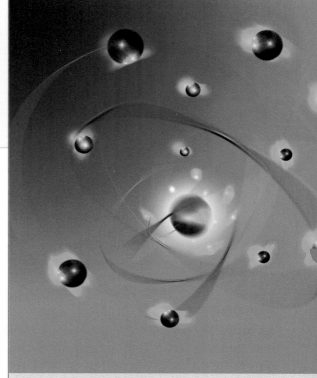

A It is difficult to imagine how electrons are arranged in shells, so a simple way to represent them is needed.

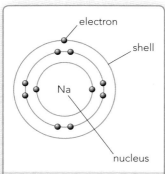

B The electronic configuration for sodium. The protons and neutrons that form the nucleus are replaced by Na, the chemical symbol for sodium.

Electrons are arranged in shells around the nucleus of the atom. Each shell is shown as a circle drawn around the chemical symbol for the atom.

The way in which the electrons are arranged in an atom is called its **electronic configuration**. Figure B shows the electronic configuration for sodium.

Figure C shows the electronic configurations for the first 20 elements in the periodic table (hydrogen to calcium).

1 What is an electronic configuration?

2 How many electrons are there in the outermost shells of lithium (Li) and sulfur (S)?

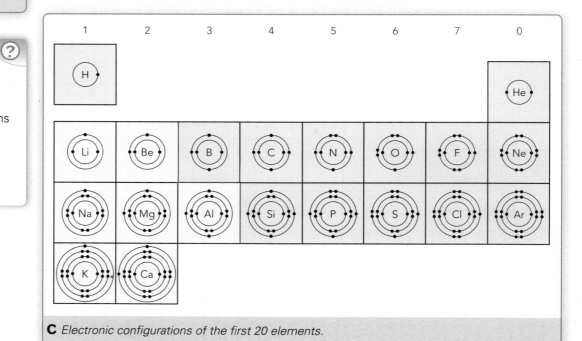

C Electronic configurations of the first 20 elements.

Finding configurations

Different shells can contain different numbers of electrons. For the first 20 elements:
- the first shell can contain up to two electrons
- the second and third shells both hold up to eight electrons (although the third shell can hold more than this).

Electronic configurations can be worked out using atomic numbers. For example, the atomic number of sodium is 11. This means that there are 11 protons and also 11 electrons. So to fill the shells:
- two electrons fit in the first shell, leaving nine
- eight electrons fit in the second shell, leaving one
- one electron fits into the third shell, leaving none.

> **3** How can you tell, from the periodic table, the number of electrons in the atoms of an element?
>
> **4** What is the maximum number of electrons allowed in each of the first two shells?

D *Electronic configuration diagrams use either dots or crosses to represent the electrons.*

Diagrams are not the only way to show electronic configuration. They can be written out instead. For sodium this is 2.8.1 – each number shows how many electrons there are in a shell. The dots separate one shell from the next. This method of showing electronic configurations is easy to write and to understand.

Spotting connections

There are several connections between an element's electronic configuration and where it is found in the periodic table. For example, the number of occupied shells is the same as the period number. Apart from the elements in group 0, which all have full outer shells, the number of outer electrons is the same as the group number. For example, the electronic configuration of sodium (2.8.1) shows that sodium is in the third period and in group 1.

> **5** Describe the connection between the electronic configuration of an element and its place in the periodic table.

> **6** Draw and write the electronic configuration for atoms with the following atomic numbers: **a** 6 **b** 17.
>
> **7** Two different elements have five outer electrons in their atoms. To which group do they belong and why?
>
> **8** The electronic configuration of an element is 2.8.7. What information does this give about the element and its atoms?

Skills spotlight

Scientists often use conventions and symbols to present information. Describe the advantages of writing electronic configurations rather than drawing them.

Learning Outcomes

1.12 Apply rules about the filling of electron shells (energy levels) to predict the electronic configurations of the first 20 elements in the periodic table, given the atomic numbers, as diagrams and in the form 2.8.1

1.13 Describe the connection between the number of outer electrons and the position of an element in the periodic table

HSW **11** Present information using scientific conventions and symbols

>>>>>>>>>>>>>>>>>>>>> Where might you find a building made of salt?

How do ionic bonds form?

A very unusual hotel in the middle of a salt lake in Bolivia, South America, has walls made entirely from blocks of common salt. Luckily there is little rain. Common salt dissolves easily in water because of its type of bonding.

A Visitors are warned not to lick the walls at the Salt Palac

B The sulfate ion SO_4^{2-} consists of a group of atoms with an overall negative charge.

An **ion** is an atom or group of atoms with a positive or negative charge. The atoms of most elements have incomplete outer shells of electrons. This means they can lose or gain electrons during chemical reactions. When this happens, the atoms become ions. Groups of atoms may also become ions if they lose or gain electrons.

Cations

Metal atoms lose electrons to form positively charged ions called **cations**. For example, sodium atoms Na become sodium ions Na^+. The + sign shows that a sodium ion carries a single positive charge.

1 What is an ion?

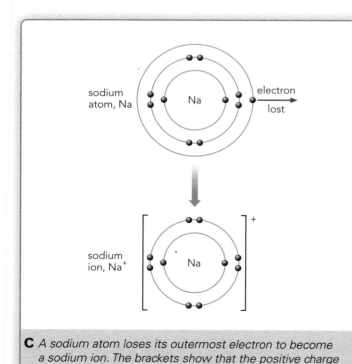

C A sodium atom loses its outermost electron to become a sodium ion. The brackets show that the positive charge belongs to the whole ion.

Group number	1	2
Number of outer electrons	1	2
Number of electrons lost	1	2
Charge on cation	1+	2+
Example	Na^+	Mg^{2+}

D In groups 1 and 2 the charge on the ion is linked to the group number.

Metal atoms lose outermost electrons when they form ions. For elements in groups 1 and 2, the number of outer electrons lost is the same as the group number.

Anions

Non-metal atoms gain electrons to form negatively charged ions called **anions**. For example, chlorine atoms Cl become chloride ions Cl^-. The − sign shows that a chloride ion carries a single negative charge.

Notice that the end of the name changes to –ide when non-metal atoms form ions.

Non-metal atoms gain electrons to complete their outermost shell when they form ions. The number of electrons that atoms in groups 6 and 7 can gain is eight minus the group number. Chlorine is in group 7, so chlorine atoms have seven outermost electrons. The outermost shell can contain eight electrons, so a chlorine atom can gain one electron to complete its outermost shell.

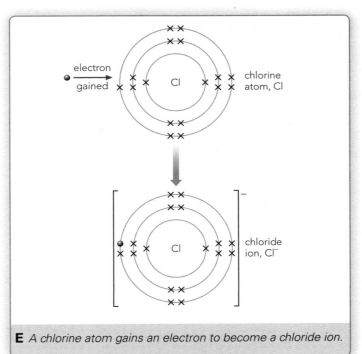

E *A chlorine atom gains an electron to become a chloride ion.*

2 Predict the symbols of the ions formed by the following metals:
a lithium (Li) found in group 1 **b** calcium (Ca) found in group 2.

3 There are three outermost electrons in an aluminium atom.
a Write down the symbol for an aluminium ion.
b Is this a cation or an anion?

4 Draw a diagram involving electron shells and electrons to show how a fluorine atom becomes a fluoride ion.

Transferring electrons

When atoms of different elements combine to form compounds, new chemical bonds form. For example, when sodium chloride forms, electrons are transferred from sodium atoms to chlorine atoms, forming Na^+ ions and Cl^- ions.
These oppositely charged ions attract each other strongly. New chemical **bonds** form, called **ionic bonds**.

Group number	6	7
Number of outer electrons	6	7
Number of electrons gained	2	1
Charge on anion	2–	1–
Example	O^{2-}	Cl^-

F *The charges on non-metal ions in groups 6 and 7.*

Watch Out!

Students are often confused about atoms and ions. Remember that an electron is negative. *Non*-metal atoms gain electrons and become *n*egative ions. Metal atoms lose electrons and become positive ions.

5 Predict the symbols of the ions formed by the following non-metals:
a sulfur (S) found in group 6 **b** iodine (I) found in group 7.

6 Explain, in terms of electrons and chemical bonds, what happens when sodium reacts with chlorine to form sodium chloride.

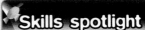

Skills spotlight

Scientists use chemical symbols for substances including ions. Describe the information contained in the symbols Mg^{2+} and O^{2-}.

Learning Outcomes

2.1 Demonstrate an understanding that atoms of different elements can combine to form compounds by the formation of new chemical bonds

2.2 Describe how ionic bonds are formed by the transfer of electrons to produce cations and anions

2.3 Describe an ion as an atom or group of atoms with a positive or negative charge

2.4 Describe the formation of sodium ions, Na^+, and chloride ions, Cl^-, and hence the formation of ions in other ionic compounds from their atoms, limited to compounds of elements in groups 1, 2, 6 and 7

HSW **11** Present information using scientific conventions and symbols

What do *sal culinaris, natrum muriaticum* and table salt all have in common?

How are the formulae of ionic compounds worked out?

The white crystals we put on chips have had many names in the past, including *natrum muriaticum*, muriate of soda, *sal culinaris* and just salt. Scientists have agreed on a name that everyone should use, sodium chloride.

A *The name on the label of this homeopathic remedy just means sodium chloride.*

1 Give the formulae for:

a sodium oxide

b magnesium chloride.

ResultsPlus
Watch Out!

Brackets must be used when more than one of the compound ions is needed to balance the charge on the other ion in a compound. For example, NaOH and $Mg(OH)_2$ are correct, but Na(OH) and $MgOH_2$ are not.

Working out a formula

Ionic compounds contain ions. You can work out the **formula** of an ionic compound if you know which ions are in it. Ionic compounds are electrically neutral because they contain equal numbers of positive and negative charges. In sodium chloride, one Na^+ ion is needed for every Cl^- ion, so its formula is NaCl.

Many ionic compounds have ions with different numbers of charges. For example, aluminium oxide contains Al^{3+} ions and O^{2-} ions. Two Al^{3+} ions give a total of six positive charges, and three O^{2-} ions give a total of six negative charges. The formula of aluminium oxide is Al_2O_3.

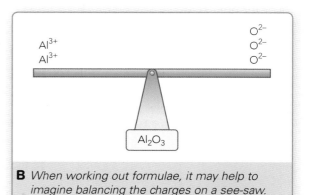

B *When working out formulae, it may help to imagine balancing the charges on a see-saw.*

Positive ions		Charge	Negative ions		Charge
ammonium	NH_4^+	1+	chloride	Cl^-	1−
potassium	K^+	1+	bromide	Br^-	1−
sodium	Na^+	1+	iodide	I^-	1−
calcium	Ca^{2+}	2+	hydroxide	OH^-	1−
magnesium	Mg^{2+}	2+	nitrate	NO_3^-	1−
copper	Cu^{2+}	2+	oxide	O^{2-}	2−
iron (II)	Fe^{2+}	2+	carbonate	CO_3^{2-}	2−
aluminium	Al^{3+}	3+	sulfate	SO_4^{2-}	2−

C *Some common ions and their formulae*

Compound ions

Compound ions contain more than one element. For example, the nitrate ion NO_3^- contains one nitrogen atom joined to three oxygen atoms and an extra electron. If two or more compound ions of the same type are needed in a formula, the ion must be written inside brackets. For example, the formula of magnesium nitrate is $Mg(NO_3)_2$. This shows that the two positive charges in one Mg^{2+} ion are balanced by two negative charges from two NO_3^- ions.

Names of ionic compounds

You may have noticed that the names of some compounds end in –ide and some end in –ate. The –ate ending shows that oxygen atoms are present.

–ide compounds		–ate compounds	
potassium bromide	KBr	potassium bromate	$KBrO_3$
iron (II) sulfide	FeS	iron sulfate	$FeSO_4$
calcium carbide	CaC_2	calcium carbonate	$CaCO_3$

D *Some examples of –ide and –ate compounds*

The structure of ionic compounds

H

The ions in an ionic compound are arranged in a regular way, called a **lattice structure**. The ions in sodium chloride pack together in a box-like arrangement, which is why it forms cube-shaped crystals. There are strong electrostatic forces of attraction between the oppositely charged ions. These are the ionic bonds.

E *The lattice structure of sodium chloride*

F *Tiny sodium chloride crystals viewed using an electron microscope (magnification x 252).*

H 5 Explain, with the help of a diagram, what is meant by lattice structure.

6 Iodised table salt contains an iodine compound to provide extra iodine, which is needed to keep us healthy. This is usually potassium iodide or potassium iodate, which contains the iodate ion IO_3^-. Describe the similarities and differences between these two compounds and give their formulae.

2 Give the formulae for:
a copper(II) sulfate
b ammonium nitrate.

3 For each of the following compounds, give the symbols of the ions it contains and its formula:
a potassium chloride
b magnesium oxide
c aluminium hydroxide
d calcium nitrate
e potassium sulfate
f ammonium carbonate.

4 Phosphide ions are P^{3-}. Name the elements contained in aluminium phosphide and aluminium phosphate.

Skills spotlight

Chemical names provide information about what the named substance contains. Table salt may contain potassium iodide or potassium iodate to avoid iodine deficiency. Explain the difference in the names of these two compounds.

Learning Outcomes

2.5 Demonstrate an understanding of the use of the endings '–ide' and '–ate' in the names of compounds

2.6 Deduce the formulae of ionic compounds (including oxides, hydroxides, halides, nitrates, carbonates and sulfates) given the formulae of the constituent ions

H 2.7 Describe the structure of ionic compounds as a lattice structure: a consisting of a regular arrangement of ions b held together by strong electrostatic forces (ionic bonds) between oppositely charged ions

HSW 11 Present information using scientific conventions and symbols

Does sodium chloride melt?

What are ionic compounds like?

Some modern bathroom scales do not just show your mass, they also show how much body fat you have. They pass a small electric current through your body. This flows through fat with difficulty but more easily through body fluids containing dissolved salts, so the proportion of body fat can be estimated.

A *Body-fat scales measure how easily electricity flows through the body.*

> **?**
>
> **1** What is an aqueous solution?
>
> **2** Sodium chloride is soluble in water, but magnesium oxide is insoluble in water. Under what conditions will:
>
> **a** sodium chloride conduct electricity
>
> **b** magnesium oxide conduct electricity?

Sodium chloride, magnesium oxide and other ionic substances do not conduct electricity when they are solid. However, they do conduct electricity when they are **molten** (heated until they turn into a liquid). This is why aluminium can be extracted from molten aluminium oxide by **electrolysis**. Ionic substances also conduct electricity when they are in **aqueous solution** (dissolved in water) – this is why sea water will conduct electricity.

H

Two conditions must be met for a substance to conduct electricity. It must contain charged particles *and* these particles must be free to move. Ionic substances contain charged particles (ions) but these are only free to move from place to place when the ionic compound is molten or in aqueous solution. In the solid form the ions are held in fixed positions, forming a lattice structure in which they can only vibrate.

> **ResultsPlus**
> **Watch Out!**
>
> Many students lose marks in exams by simply stating that ionic compounds can conduct electricity. You need to remember the conditions under which they will conduct electricity and that the ions will have to move.

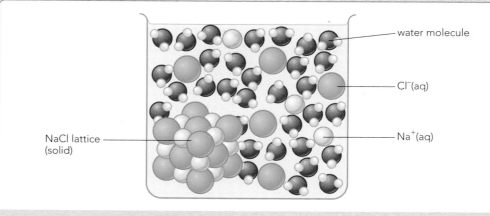

NaCl lattice (solid)

water molecule

Cl⁻(aq)

Na⁺(aq)

B *Ions are free to move in aqueous ionic substances.*

> **?**
>
> **3** Explain the ability of ionic compounds to conduct electricity when:
>
> **a** molten
>
> **b** in aqueous solution.

Melting points and boiling points

The **melting point** of a substance is the temperature at which it changes from a solid to a liquid. Its **boiling point** is the temperature at which it changes from a liquid to gas at its fastest possible rate. Ionic substances have high melting points. They also have high boiling points. This means that they are usually solids at room temperature.

Ionic substance	Melting point in °C	Boiling point in °C
sodium chloride, NaCl	801	1413
magnesium oxide, MgO	2852	3600

C *Sodium chloride and magnesium oxide have high melting and boiling points.*

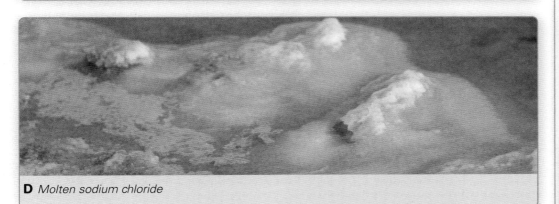

D *Molten sodium chloride*

Ionic bonds are strong. Energy is needed to break an ionic bond, and there are very many of them in an ionic compound. This is why ionic compounds have high melting points and boiling points. **H**

5 Substance X melts at 1530 °C and conducts electricity when solid. Explain why it is not an ionic compound.

6 Substance Y melts at 373 °C. Describe a simple test you could do to see if it is an ionic compound.

H 7 a Explain why ionic compounds are usually solid at room temperature.
b Using the data on this page, suggest which compound – sodium chloride or magnesium oxide – is likely to have the stronger ionic bonds. Explain your answer.

8 Write bullet-pointed revision notes for the properties of ionic compounds. Include melting points, boiling points and conduction of electricity.

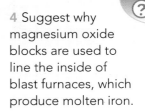

4 Suggest why magnesium oxide blocks are used to line the inside of blast furnaces, which produce molten iron.

Skills spotlight

The use of science and technology can lead to significant benefits. One type of industrial electrical cable, called MICC, contains several copper wires inside a protective copper tube. Magnesium oxide powder, an electrical insulator, is packed between the wires.

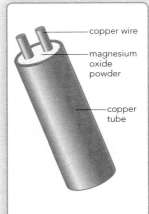

E *Cross-section through a MICC cable*

Suggest an advantage of this type of industrial cable over ordinary plastic-coated cable in the event of a fire.

Learning Outcomes

2.8 Describe **H** and explain the properties of ionic substances including sodium chloride and magnesium oxide, limited to:
a melting points and boiling points **b** their ability to conduct electricity as solids, when molten and in aqueous solution

HSW 12 Describe the benefits, drawbacks and risks of using new scientific and technological developments

C2.8 Solubility

Is it possible to predict whether a salt will be soluble or insoluble?

Aspirin, acetylsalicylic acid, is a household medicine used to ease pain. Aspirin tablets are not very soluble. They have to be swallowed with a glass of water and it takes time to be absorbed by the stomach. To get around these problems, medicine manufacturers have made aspirin into various soluble substances. These are absorbed much faster, providing quicker relief from pain.

A *Some aspirin medicines dissolve better than others.*

1 What is the difference between a substance that is soluble in water and one that is insoluble?

If a substance **dissolves** well in a particular liquid, we say that it is **soluble**. If a substance does not dissolve at all in a particular liquid, we say that it is **insoluble**. A **salt** is a substance that can be made by reacting an acid with an alkali. Figure C shows some rules for identifying which salts are soluble in water and which are not.

B *Fertilisers contain soluble salts that plants can absorb through their roots.*

Soluble in water	Insoluble in water
all common sodium, potassium and ammonium salts	
all nitrates	
most chlorides	silver chloride, lead chloride
most sulfates	lead sulfate, barium sulfate, calcium sulfate
sodium carbonate, potassium carbonate, ammonium carbonate	most carbonates
sodium hydroxide, potassium hydroxide, ammonium hydroxide	most hydroxides

C *Some soluble and insoluble salts*

2 Which substances in this list are soluble in water and which are insoluble in water? Sodium chloride, lead nitrate, calcium sulfate, potassium hydroxide, silver chloride, calcium carbonate, ammonium carbonate.

Precipitates

Lead nitrate and potassium iodide are salts that are soluble in water. Their solutions react together to produce lead iodide and potassium nitrate. Lead iodide is insoluble in water. It appears as a yellow **precipitate**, made up of particles of lead iodide suspended in the liquid. Here are the equations for the reaction:

lead nitrate + potassium iodide → lead iodide + potassium nitrate

$$Pb(NO_3)_2(aq) + 2KI(aq) \rightarrow PbI_2(s) + 2KNO_3(aq)$$

Notice how the equation is correctly balanced using numbers in front of KI and KNO_3. The **state symbols** show that all the substances are dissolved in water, apart from lead iodide, which is insoluble and so is present as a solid. A reaction in which an insoluble solid is produced from two soluble substances is called a **precipitation reaction**. It happens if the reactants are soluble and at least one of the products is insoluble but not if all the products are soluble.

D Lead iodide forms a yellow precipitate.

ResultsPlus
Watch Out!

The word precipitate can be used to describe a chemical formed in a precipitation reaction as well as what it does. It can precipitate to the bottom of the beaker. As in 'The yellow precipitate of lead iodide forms and precipitates to the bottom of the beaker.'

3 What is a precipitate?

4 How can you tell from the equation on page 132 that lead iodide forms a precipitate?

5 Explain why lead chloride could form a precipitate but potassium chloride could not.

6 Potassium sulfate K_2SO_4 and barium nitrate $Ba(NO_3)_2$ are soluble in water. They react to form potassium nitrate KNO_3 solution and insoluble barium sulfate $BaSO_4$.
a Write a word equation for the reaction.
b Write a balanced equation for the reaction and include the state symbols.

7 A school laboratory technician has been asked to make solutions containing a chloride, a nitrate, a carbonate, a sulfate, a hydroxide and a silver salt. Suggest, with reasons, suitable substances that the technician could use.

Skills spotlight

Scientists may communicate their work in a quantitative way – which involves numbers. They may also communicate their work in a qualitative way – which involves giving observations without the use of numbers. Give one example from this spread of a quantitative description and one example of a qualitative description.

Learning Outcomes

2.9 Recall the general rules which describe the solubility of common types of substances in water: a all common sodium, potassium and ammonium salts are soluble b all nitrates are soluble c common chlorides are soluble except those of silver and lead d common sulfates are soluble except those of lead, barium and calcium e common carbonates and hydroxides are insoluble except those of sodium, potassium and ammonium

2.10 Demonstrate an understanding that insoluble salts can be formed as precipitates by the reaction of suitable reagents in solution

HSW 10 Use qualitative and quantitative approaches when presenting scientific ideas and arguments, and recording observations

>>>>>>>>>>>>>>>>>>>>>>>> How do rock climbers stop their hands getting sweaty on a cliff face?

C2.9 Precipitation

How can we make magnesium carbonate using a precipitation reaction?

Rock climbing can be scary, putting you under stress and making you sweat. You do not want sweaty, slippery hands when you are climbing a rock face several hundred metres above the ground. Rock climbers rub 'chalk' into their hands to keep them dry. This is not the sedimentary rock made from calcium carbonate. Instead, it is magnesium carbonate, $MgCO_3$, which is able to absorb water really well. Without it, rock climbing would be even scarier.

Some salts are soluble in water but others are not. For example, all sodium salts are soluble and most sulfates are soluble. Some combinations of soluble salts will react together to form an insoluble salt, the precipitate. For example, silver nitrate and sodium chloride are soluble. Their solutions react together, when mixed, to form a precipitate of silver chloride:

A This climber is on a sheer rock face. Magnesium carbonate is 'hygroscopic', which means that it absorbs moisture. This property helps to keep the climber's hands dry, providing more friction between them and the rock.

$$\text{silver nitrate} + \text{sodium chloride} \rightarrow \text{silver chloride} + \text{sodium nitrate}$$
$$AgNO_3(aq) + NaCl(aq) \rightarrow AgCl(s) + NaNO_3(aq)$$

Notice how the ions involved appear to have 'swapped places' in this precipitation reaction. Silver ions from the silver nitrate solution combine with chloride ions from the sodium chloride solution to form insoluble silver chloride. This precipitate can then be separated from the unreacted ions in solution by **filtration**, washed on the filter paper, then dried in a warm oven.

Your task

You are going to plan an investigation to test how the amount of magnesium carbonate formed in a precipitation reaction changes when one of the substrates is varied. Your teacher will provide you with some materials to help you organise this task.

Learning Outcomes

2.11 Demonstrate an understanding of the method needed to prepare a pure, dry sample of an insoluble salt

2.12 Prepare an insoluble salt by precipitation

Build Better Answers

When planning an investigation like this, you will be assessed on your *overall plan*. There are 4 marks available for this skill. Here are two extracts that focus on this skill. Other skills that you need for the assessment are dealt with in other lessons.

Student extract 1 — A basic response for this skill

This is a good start because the student has given clear instructions on how they are planning to carry out the practical.

> I am going to add 25 cm³ of sodium carbonate solution to a beaker and then add 10 cm³ of magnesium sulfate solution. This will form magnesium carbonate, which is insoluble. I will record the mass of a piece of filter paper and then use the filter paper to filter the magnesium carbonate out of the solution. I will wash the paper and residue with water and then dry it in a warm oven. I will then measure the mass of the filter paper and the dry magnesium carbonate. I will then do this experiment again with different volumes of magnesium nitrate solution.

However, they will not be able to access all the marks because they do not explain the range of other volumes they are planning to test.

Student extract 2 — A good response for this skill

Here, the student continues from the end of the extract above.

The method explains how the hypothesis will be tested.

> … I will try it again with 10 cm³, 15 cm³, 20 cm³ and 25 cm³ of the magnesium sulfate solution. I have chosen this range because it will allow me to work out if less or the same amount of each solution is needed to get the maximum amount of precipitate. I think that by carrying out this investigation I will be able to test my hypothesis that the amount of magnesium carbonate formed depends on the volume of magnesium sulfate used.

The student has stated the range of tests they are going to carry out.

The student has also explained why this range has been chosen.

 ResultsPlus

To access 2 marks

- Provide a logically ordered method that will produce results
- Choose a range of data or observations that will test the hypothesis

To access 4 marks

You also need to:
- Explain how your method will test the hypothesis
- Explain why you have chosen your range of data or observations

C2.10 Precipitates

Can we predict if a precipitate will form in a chemical reaction?

Before the photocopier was invented in the 1930s, diagrams were copied by blueprinting. Paper was treated with a mixture of two soluble iron salts. The original diagram, on tracing paper, was laid on top and left in sunlight. A blue precipitate formed in the lit areas, leaving white lines on a blue background.

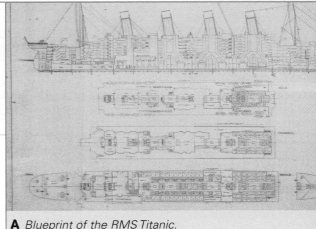

A *Blueprint of the RMS Titanic.*

B *Copper hydroxide forms a pale blue precipitate.*

It is possible to work out which substances are needed to make a particular precipitate, and also to work out what precipitate will form when two solutions are mixed together.

Consider the precipitation reaction studied on the previous page:

lead nitrate	+	potassium iodide	→	lead iodide	+	potassium nitrate
(soluble)	+	(soluble)	→	(insoluble)	+	(soluble)

The ions from the soluble salts have swapped:
- lead iodide forms when lead ions and iodide ions meet
- potassium nitrate forms when potassium ions and nitrate ions meet.

All nitrates are soluble, so potassium nitrate does not form a precipitate. However, most lead salts are insoluble, so lead iodide forms a precipitate. Figure C on page 132 helps you to predict the formation of a precipitate when two named solutions are mixed together – for example, when solutions of copper chloride and potassium hydroxide are mixed:

copper chloride	+	potassium hydroxide	→	copper hydroxide	+	potassium chloride
$CuCl_2$	+	$2KOH$	→	$Cu(OH)_2$	+	$2KCl$ (soluble)

All common potassium salts and most chlorides are soluble. So potassium chloride does not form a precipitate. Most hydroxides are insoluble according to Figure C on page 132, so copper hydroxide is likely to form a precipitate.

> **1** Copy the balanced equation and add the correct state symbols. ?

> ?
> **2** A precipitation reaction occurs when calcium chloride solution and potassium sulfate solution are mixed together.
> **a** Write a word equation for the reaction.
> **b** Using Figure C on page 132, predict which of the two products will form a precipitate.

Barium meals

Barium sulfate is given to patients to help in diagnosing problems with their intestines. Like bone, barium sulfate is opaque to X-rays, which means that X-rays do not pass through it and are absorbed. The patient swallows a drink, called a **barium meal**, which contains barium sulfate. X-ray photographs are then taken as the barium sulfate passes through the patient's digestive system. The barium sulfate shows up as white on X-ray photographs.

Barium sulfate is used for barium meals but most barium salts are **toxic**. However, barium sulfate is insoluble and this prevents it from entering the blood. This makes it safe to swallow.

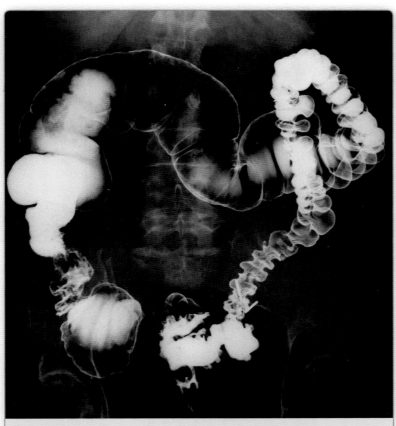

C *Barium sulfate shows up as white on X-ray photographs of the digestive system.*

3 Barium chloride $BaCl_2$ and ammonium sulfate $(NH_4)_2SO_4$ react together in a precipitation reaction, forming barium sulfate $BaSO_4$ and ammonium chloride NH_4Cl.
a Write a word equation for this reaction.
b Write a balanced equation for the reaction and include state symbols.

ResultsPlus
Watch Out!

It is useful to remember that to make an insoluble salt XY, mixing X nitrate solution and sodium Y solution will always work.

4 A precipitation reaction occurs when silver nitrate solution and sodium chloride solution are mixed together.
a Write a word equation for the reaction.
b Using Figure C on page 132, predict which of the two products will form a precipitate.

5 Lead carbonate is insoluble. Identify two pairs of solutions that would each make lead carbonate, giving reasons for your choices. Write a word equation for each reaction.

Skills spotlight

It is important that scientists test their ideas. Explain how it is possible to predict whether or not a precipitate will form when two solutions are mixed, and describe how you would test this hypothesis.

Learning Outcomes

2.13 Use solubility rules to predict whether a precipitate is formed when named solutions are mixed together and to name the precipitate.

2.14 Recall that the insoluble salt, barium sulfate, is given as a 'barium meal' to X-ray patients because
a it is opaque to X-rays **b** it is safe to use as, although barium salts are toxic, its insolubility prevents it entering the blood

HSW **5** Plan to test a scientific idea, answer a scientific question or solve a scientific problem

>>>>>>>>>>>>>>>>>>>>>>>>>>>>>> What produces a blue-green flame when it is burned?

How do scientists identify the ions in a substance?

Different metals produce different colours when they are put in a hot flame (see Figure A). This is the basis of atomic absorption spectroscopy, used to identify the elements in samples of water.

A

1 Describe how to test a substance for the presence of sodium ions.

2 Explain why Bunsen and Kirchhoff concluded from their experiments that caesium and rubidium were two different elements.

Flame tests are carried out on samples of ionic substances. The most intense colours are obtained from solids, but flame tests also work when the solids are dissolved in water.

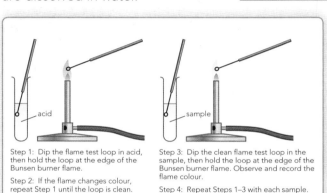

Step 1: Dip the flame test loop in acid, then hold the loop at the edge of the Bunsen burner flame.

Step 2: If the flame changes colour, repeat Step 1 until the loop is clean.

Step 3: Dip the clean flame test loop in the sample, then hold the loop at the edge of the Bunsen burner flame. Observe and record the flame colour.

Step 4: Repeat Steps 1–3 with each sample.

B How to carry out a flame test.

Metal ion		Flame test colour
sodium	Na$^+$	yellow
potassium	K$^+$	lilac
calcium	Ca^{2+}	red
copper(II)	Cu^{2+}	green-blue

C Flame test colours produced by four different metal ions.

D The flame test for caesium

E The flame test for rubidium

Flame tests have led to the discovery of new elements. In 1860, Robert Bunsen (1811–1899) and Gustav Kirchhoff (1824–1887) used simple **spectroscopy** to study the light given off in flame tests by some samples of mineral water. They used a prism to separate the colours in the light into a spectrum and saw a grey-blue colour that had not been seen before. They realised that they had discovered a new element, which they called caesium, after the Latin word for sky-blue. A year later using the same technique they discovered another new element. They named it rubidium, after the Latin word for dark red.

Precipitation tests

Certain anions can be identified by precipitation tests. To identify aqueous chloride ions, Cl^-(aq), a few drops of dilute nitric acid are added to the sample solution, which is then shaken. A few drops of silver nitrate solution are then added. A white precipitate of silver chloride forms if the sample contains chloride ions.

To identify aqueous sulfate ions, SO_4^{2-}(aq), a few drops of dilute hydrochloric acid are added to the sample solution, which is then shaken. A few drops of barium chloride solution are then added. A white precipitate of barium sulfate forms if the sample contains sulfate ions.

Test for carbonate ions

Carbon dioxide is given off when an acid is added to a substance containing carbonate ions, CO_3^{2-}. Limewater is used to confirm the identity of the gas. Limewater turns milky when carbon dioxide is bubbled through it.

F Silver chloride precipitates in the test for Cl^-.

G Barium sulfate forms a white precipitate in the test for SO_4^{2-}

3 Hydrochloric acid contains chloride ions. Explain why hydrochloric acid cannot be used instead of nitric acid in the silver nitrate test.

4 Sulfuric acid contains sulfate ions. Explain why sulfuric acid cannot be used instead of hydrochloric acid in the barium chloride test.

5 A white solid produces a lilac flame in a flame test. It also produces a white precipitate when tested with barium chloride solution but not when tested with silver nitrate solution.
a Name the metal cation present.
b Name the anion present.
c Give the name of the white solid.

6 A light green solid produces a green-blue flame in a flame test. It also produces a white precipitate when tested with silver nitrate solution but not when tested with barium chloride solution. Give the name and formula of the light green solid.

7 Describe how the laboratory tests described here could be used to distinguish between sodium chloride, potassium chloride, calcium sulfate and sodium carbonate.

ResultsPlus
Watch Out!

Take care not to confuse the acids needed for precipitation tests. *Nitric acid goes with silver nitrate*, and *hydrochloric acid goes with barium chloride*.

Learning Outcomes

2.15 Describe tests to show the following ions are present in solids or solutions: a Na^+, K^+, Ca^{2+}, Cu^{2+} using flame tests b CO_3^{2-} using dilute acid and identifying the carbon dioxide evolved c SO_4^{2-} using dilute hydrochloric acid and barium chloride solution d Cl^- using dilute nitric acid and silver nitrate solution

2.16 Recall that chemists use spectroscopy to detect the presence of very small amounts of elements and that this led to the discovery of new elements, including rubidium and caesium.

HSW **2** Describe how data is used by scientists to provide evidence that increases our scientific understanding

How are atoms of non-metals held together?

Theobromine is one of the molecules in chocolate that makes it taste good. Now you can buy necklaces that show a model of a theobromine molecule. The straight parts of the model show the links between atoms.

A Jewellery inspired by chocolate.

There are many different kinds of molecule in your body, and most of the atoms making up these molecules are non-metal atoms. Most non-metal compounds are held together by pairs of electrons, which form **covalent bonds**. The electrons used in these bonds come from the outermost electron shells of the atoms.

Covalent bonds can form between atoms of the same element. For example, in hydrogen gas two hydrogen **atoms** form one hydrogen **molecule**. Each atom contributes one electron to the covalent bond, allowing both atoms to have full outermost electron shells (remember that the first shell can only have two electrons in it – see page 125).

Figure B is a **dot and cross diagram** showing a hydrogen molecule. Dots and crosses are used to represent the electrons so that you can see which atom each electron is from. There is no difference between the electrons from different atoms.

Covalent bonds also form between atoms of different non-metal elements because all these elements are short of having a full outermost shell of electrons. Figure C shows three other covalent molecules.

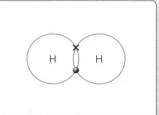

B A dot and cross diagram of a hydrogen molecule (H_2)

1 a How many electrons does a hydrogen atom have?
b How many electrons does hydrogen need to have a full outermost shell?

2 a What is a covalent bond?
b Why do non-metal atoms form covalent bonds?

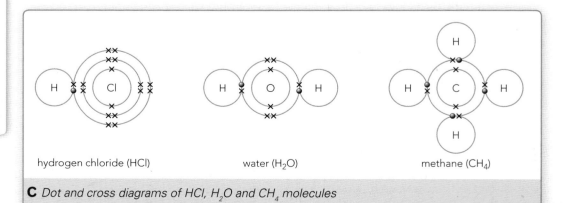

hydrogen chloride (HCl)   water (H_2O)   methane (CH_4)

C Dot and cross diagrams of HCl, H_2O and CH_4 molecules

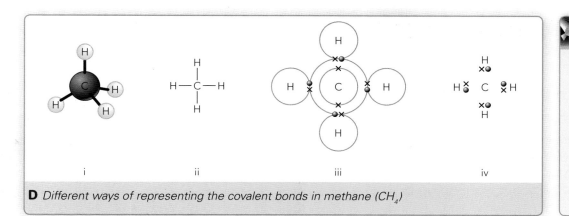

D *Different ways of representing the covalent bonds in methane (CH₄)*

Skills spotlight

Scientists use different conventions to represent molecules. Figure D shows some different ways of representing the bonds in a molecule of methane. What are the advantages and disadvantages of each type of drawing?

Atoms can share more than one pair of electrons if this is needed for each atom to have a full outermost shell. Two pairs of shared electrons form a **double bond**.

3 Look at Figure F and compare it to part iii of Figure D. Suggest why the diagram does not need to show the first electron shell in the carbon atom.

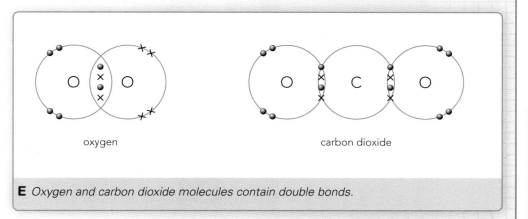

E *Oxygen and carbon dioxide molecules contain double bonds.*

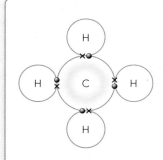

F *Dot and cross diagram of methane without the innermost electron shell of the carbon atom.*

4 The electronic configuration of carbon is 2.4. How does this tell you that carbon atoms need to share four electrons with other atoms to get a full outermost shell?

5 The electronic configuration of oxygen is 2.6. Draw a diagram to help you to explain why one oxygen atom forms covalent bonds with *two* hydrogen atoms to form a water molecule.

6 Describe what a covalent bond is and why carbon forms covalent bonds with four hydrogen atoms to form a methane molecule.

Learning Outcomes

3.1 Describe a covalent bond as a pair of electrons shared between two atoms

3.2 Recall that covalent bonding results in the formation of molecules

3.3 Explain the formation of simple molecular, covalent substances using dot and cross diagrams, including:
   a hydrogen **b** hydrogen chloride **c** water **d** methane **H** **e** oxygen **f** carbon dioxide

HSW **1** Present information using scientific conventions and symbols

>>>>>>>>>>>>>> Sand and oxygen both contain covalent bonds. Why is one a solid and one a gas?

What are the properties of covalent substances?

This skull was created by artist Damien Hirst – it is decorated with 221 g of diamonds worth £15 000 000. Diamonds are a form of carbon but the same mass of normal, black-powdered carbon is worth only about £1.

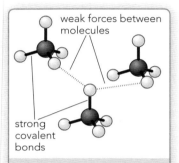

B *There are only weak forces between neighbouring molecules of methane.*

weak forces between molecules

strong covalent bonds

Simple molecular covalent substances

Simple molecular covalent substances include gases such as hydrogen, methane, oxygen and carbon dioxide, and liquids such as water. These substances have low melting and boiling points because there are only weak forces between neighbouring molecules.

An electric current is a flow of charged particles. The atoms in simple molecular covalent molecules have not lost or gained electrons, so there are no charged particles that can move around. These substances are poor conductors of electricity.

Giant molecular covalent substances

Some covalent compounds form **giant molecular covalent substances**. These consist of billions of atoms all joined together by covalent bonds. Most giant molecular substances have high melting and boiling points. This is because all the atoms are joined to other atoms by strong covalent bonds, and a lot of heat energy is needed to break these bonds.

Sand is made of silicon and oxygen atoms joined together in a giant molecular structure. Sand grains are very hard. Carbon can form two different types of giant molecular structure: **diamond** which is very hard and **graphite** which is very soft.

A *Damien Hirst created this diamond encrusted skull.*

C *Diamond will scratch most other substances, and it is used to cut patterns into glass.*

1 List three compounds that are simple molecular covalent substances.

2 a Why do most simple molecular covalent substances have low melting and boiling points?
b Why don't they conduct electricity?

3 a Name three giant molecular covalent substances.
b Compare the melting points of these substances to those you named in question 1.
c Explain why the melting points are different.

Diamond and graphite have very different properties because of the different ways in which the carbon atoms are arranged. Both have very high melting and boiling points because of the strong covalent bonds between the carbon atoms.

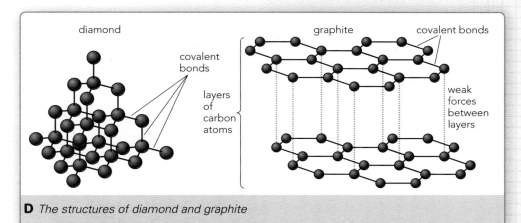

D *The structures of diamond and graphite*

There are also differences in the properties.

Diamond:
- is very hard because all the atoms are joined with strong covalent bonds. Diamonds are used to make cutting tools.
- does not conduct electricity because there are no free electrons or charged particles to move around.

Graphite:
- easily rubs away in layers because although the covalent bonds within the layers are very strong, there are only weak forces *between* the layers. It is soft enough to be used as a **lubricant**.
- conducts electricity because there is one electron from each carbon atom that can move along the layers. This means that graphite can be used to make electrodes.

H 5 What property of graphite makes it an unusual giant molecular covalent compound?

H 6 a Explain why diamond is typical of giant molecular covalent substances.
b Explain why graphite is not a typical example.

7 Describe the differences between methane and diamond and **H** explain the differences in terms of their structures.

ResultsPlus
Watch Out!

Simple molecular, covalent substances melt and boil at lower temperatures than giant molecular, covalent ones. Remember that this is because the weak forces *between* the molecules are overcome, not the strong covalent bonds *in* the molecule.

H 4 Why are the substances you named in question **3a** usually very hard?

Skills spotlight

Scientists often explain observations using models. How does the model for covalent bonding explain why liquid methane will not conduct electricity?

Learning Outcomes

3.5 Describe the properties of typical simple molecular, covalent compounds, limited to: **a** low melting points and boiling points, in terms of weak forces between molecules **b** poor conduction of electricity

3.6 Demonstrate an understanding of the differences between the properties of simple molecular, covalent substances and those of giant molecular, covalent substances, including diamond and graphite

H 3.7 Explain why, although they are both forms of carbon and giant molecular substances, graphite is used to make electrodes and as a lubricant, whereas diamond is used in making cutting tools

HSW 3 Describe how phenomena are explained using scientific models

C2.14 Classifying substances

How can you use the properties of a substance to work out what kind of bonding it has?

Industrial accidents and spillages can lead to unknown substances escaping into the environment. This can lead to serious problems. Scientists need to test these substances to identify them. Once they know what the substance is, a plan can be made to safely deal with the situation.

Almost all the substances around you consist of atoms joined together by chemical bonds. Substances containing only non-metal elements, or those containing both non-metal and metal atoms, can be joined by ionic or covalent bonds. These different types of bonding give substances different properties.

This also means that you can use the properties of a substance to help you to work out what kind of bonding holds the atoms in the substance together.

A *A scientist testing a sample of water.*

Your task

You are going to plan an investigation that will allow you to classify different types of compounds by investigating some of their properties. You will be given the following substances: sodium chloride, magnesium sulfate, hexane, liquid paraffin, silicon dioxide, copper sulfate, and sucrose (sugar). Your teacher will provide you with some other materials to help you organise this task.

Learning Outcomes

3.4 Classify different types of elements and compounds by investigating their melting points and boiling points, solubility in water and electrical conductivity (as solids and in solution) including sodium chloride, magnesium sulfate, hexane, liquid paraffin, silicon dioxide, copper sulfate, and sucrose (sugar)

When planning an investigation like this, one of the skills you will be assessed on is your ability to *evaluate your conclusion.* There are 4 marks available for this skill. Here are two extracts focusing on this skill. Other skills that you need for the assessment are dealt with in other lessons.

Student extract 1 — A basic response for this skill

The small number of substances tested is a weakness, not a strength.

> My conclusion is good because it is correct for the few substances I tested. I could make my conclusion better by testing some more substances.

The student needs to say *why* they think the conclusion is correct.

The student comments on how the conclusion could be strengthened.

Student extract 2 — A good response for this skill

The student has pointed out a limitation in their conclusion.

They have suggested how to extend the investigation.

> The 'rules' I worked out for identifying the type of bonding work for all the substances I tested. However there may be other substances that do not fit these 'rules', so my conclusion may not be true for all substances. The results I gathered did fit with the scientific knowledge I have about how the bonding in a molecule affects it's properties.
> I should test a greater range of substances, and I could find out if there are any other physical properties I could use to help me to work out the type of bonding.

For higher marks it is important to evaluate your conclusion using scientific knowledge as well as other evidence.

 ResultsPlus

To access 2 marks

- Evaluate how well all your evidence supports your conclusion
- Suggest how your evidence can be improved to strengthen your conclusion

To access 4 marks

You also need to:
- Evaluate how well other scientific ideas support your conclusion
- Suggest how the investigation could be extended to support your conclusion

How do you separate mixtures of liquids?

This bird has been covered in oil from an oil spill. The oil won't dissolve in water and it stops the bird from flying. The oil needs to be removed from the feathers to save the bird's life.

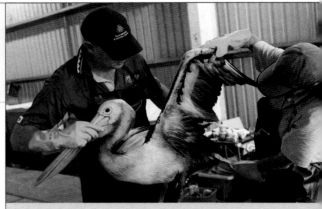

A *One of hundreds of oiled birds being cleaned following the Deepwater Horizon oil rig explosion in 2010.*

Immiscible liquids

Oils do not dissolve in water but form a separate layer floating on top. Liquids that do not mix completely with each other are **immiscible**. When you shake up immiscible liquids they seem to mix but soon separate out again.

Immiscible liquids can be separated using a **separating funnel**. When you open the tap, the lower liquid runs out of the funnel and can be collected in a beaker. You then close the tap before the other liquid starts to run out. Put another beaker under the funnel then open the tap to allow the upper liquid to be collected.

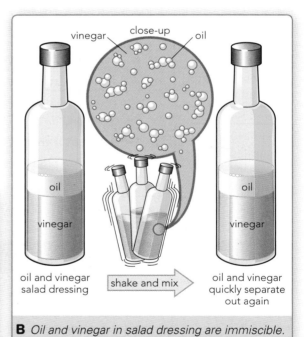

B *Oil and vinegar in salad dressing are immiscible.*

C *A separating funnel*

1 Why do oil slicks float on the sea?

2 Describe how you would use a separating funnel to separate oil and water.

Miscible liquids

If two liquids **dissolve** in each other, their particles mix completely to make a **solution**. Liquids like this are **miscible**. They do not separate out once mixed and can only be separated by **fractional distillation**. Fractional distillation can be used to separate mixtures of miscible liquids because they have different boiling points.

The mixture of liquids is heated and the liquids evaporate. The vapours condense in a **fractionating column**. The **fraction** with the highest boiling point condenses near the bottom of the column, and the fraction with the lowest boiling point reaches the top of the column.

This process is used to separate the gases in air. However, first the air has to be turned into a mixture of liquids. This is done by cooling the air so that the water can freeze and be removed. Then the remaining air is cooled to −200 °C when it is **liquefied**.

Water	freezes at 0 °C
Carbon dioxide	freezes at −79 °C
Oxygen	boils at −183 °C
Nitrogen	boils at −196 °C

D *Freezing and boiling points of the gases in air*

In the fractionating column, liquid air is warmed up to −185 °C. This causes all the nitrogen to evaporate and rise up the column. Most of the oxygen stays as a liquid and is piped out of the bottom of the column. The oxygen that does evaporate condenses again at the top of the column and runs back down to the bottom.

3 When air is cooled, which substance will solidify first?

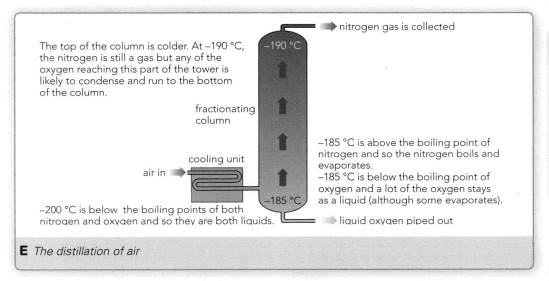

The top of the column is colder. At −190 °C, the nitrogen is still a gas but any of the oxygen reaching this part of the tower is likely to condense and run to the bottom of the column.

nitrogen gas is collected

−190 °C

fractionating column

−185 °C is above the boiling point of nitrogen and so the nitrogen boils and evaporates.
−185 °C is below the boiling point of oxygen and a lot of the oxygen stays as a liquid (although some evaporates).

cooling unit

air in

−185 °C

−200 °C is below the boiling points of both nitrogen and oxygen and so they are both liquids.

 liquid oxygen piped out

E *The distillation of air*

Skills spotlight

The oxygen collected from the fractionating column contains some argon. The boiling point of argon is −186 °C. How could you separate oxygen from argon in a second fractionating column?

4 The top of the fractionating column is at about −190 °C. Is that above or below the boiling temperature of oxygen?

5 Hexane boils at 69 °C and octane boils at 126 °C.
a Suggest what temperatures would be needed at the top and bottom of a fractionating column to separate a mixture of hexane and octane.
b Which substance would be collected from the top of the fractionating column?

6 Explain how we can separate oil from vinegar and oxygen from air. Make sure that you include the words fractional distillation, immiscible, miscible and separating funnel in your answer. ✏

ResultsPlus
Watch Out!

Many students lose marks by getting confused with negative temperatures. A liquid with a boiling temperature of −100 °C boils at a *lower* temperature than one with a boiling temperature of −50 °C.

Learning Outcomes

3.8 Describe the separation of two immiscible liquids using a separating funnel

3.9 Describe the separation of mixtures of miscible liquids by fractional distillation, by referring to the fractional distillation of liquid air to produce nitrogen and oxygen

HSW **5** Plan to test a scientific idea, answer a scientific question or solve a scientific problem

〉〉〉〉 In 2005 £100m was spent destroying food containing a banned substance. How was the substance found? 147

What is chromatography and when is it useful?

In 1983, the Sunday Times newspaper paid over £500 000 (in today's money) to publish diaries written by Adolf Hilter. However, scientists analysed the inks in the diaries and discovered that the inks were not available during Hitler's life – the diaries were forgeries.

A *One of the fake diaries*

1 How many different compounds are in the colour of substance X in Figure B?

Inks, paints and foods often contain mixtures of coloured compounds. **Chromatography** can be used to find out which colourings are in those substances.

Some coloured compounds dissolve better in a **solvent** than others. They have different **solubilities**. This difference is used to separate different components of mixtures in chromatography.

In paper chromatography, samples are placed near the bottom of a sheet of special paper. Solvent then soaks up the paper. The different components of a mixture are carried by the solvent at different speeds so they are separated. The paper with the separated components on it is called a **chromatogram**.

Results Plus
Watch Out!

A compound never rises as fast as the solvent, so R_f values are always less than 1. If you calculated an R_f value bigger than 1, you've made a mistake.

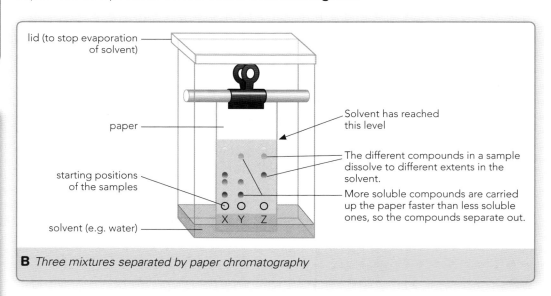

lid (to stop evaporation of solvent)

paper

Solvent has reached this level

The different compounds in a sample dissolve to different extents in the solvent.

starting positions of the samples

More soluble compounds are carried up the paper faster than less soluble ones, so the compounds separate out.

solvent (e.g. water)

X Y Z

B *Three mixtures separated by paper chromatography*

2 Look at Figure B.
a How many coloured compounds are in mixture X?
b For mixture Y, explain why the green spot is higher than the red spot.
c The solvent has moved 10 cm and the pink compound in Y has moved 2.5 cm. What is the R_f value of this compound?
d State one variable that might change a compound's R_f value.

The R_f **value** is the distance the compound has risen divided by the distance the solvent has risen.

$$R_f = \frac{\text{distance moved by compound}}{\text{distance moved by solvent}}$$

The R_f value of a particular compound does not change if the chromatography conditions are the same. R_f values can be used to identify compounds.

Uses of chromatography

Scientists at the Food Standards Agency use chromatography to separate and identify food colourings. This helps to keep our food safe.

The police use chromatography to analyse DNA samples to work out if suspects have been at a crime scene. Chromatography can also be used to analyse paints and dyes. This helps museum staff to mix exact copies of old-fashioned paints, to restore old paintings or to identify forgeries.

C *Sudan-1 is a banned food colouring. In 2005, it was discovered in some chilli powder. Foods made using this chilli powder were all destroyed.*

3 In Figure D, the chromatogram on the left shows some common colouring agents found in sweets. The chromatogram on the right shows results for some sweets.

E104 E110 E120 E122 E133

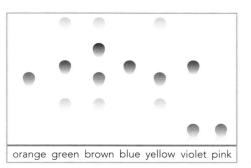

orange green brown blue yellow violet pink

D *Chromatograms showing an analysis of colourings in sweets*

a Which sweets contain one colouring agent?
b Which colouring agents are in the yellow sweet?
c Which colouring agents are in the green sweet?
d The solvent reached a height of 15 cm and the blue spot reached a height of 3 cm. What is the R_f value of the blue colouring agent?
e Why must both chromatograms be done in identical conditions?
f What is the colour of the most soluble substance in these chromatograms?

4 Describe how chromatography can help keep our food safe.

5 Police have taken four orange lipsticks from suspects. Explain the steps needed to show that one of the lipsticks made a mark at a crime scene.

Skills spotlight

There are some questions that science cannot tackle. Think up a question about a painting that can be answered by science and one that science cannot address.

ResultsPlus
Watch Out!

If you draw the apparatus for chromatography, make sure that the line for the solvent is above the bottom edge of the paper but below the samples. This means the solvent will rise to the samples. If you draw it above the samples they will just wash into the solvent.

Learning Outcomes

3.10 Describe how paper chromatography can be used to separate and identify components of mixtures, including colouring agents in foodstuffs

3.11 Evaluate the information provided by paper chromatograms, including the calculation of R_f values, in a variety of contexts, such as the food industry and forensic science

HSW 4 Identify questions that science cannot address and explain why these questions cannot be addressed.

>>>>>>>>>>>>>>>>>>>>>>>>>> In how many different ways can atoms be held together?

What are the different types of chemical bonds?

The photo shows part of an artwork called *Another Place* by Antony Gormley. It consists of 100 human figures made of iron, spread out along two miles of beach. Most of them are underwater when the tide is in. The substances in the air, water, sand and iron contain different types of bonding between their atoms.

A *Another Place*

The **atoms** of **elements** in a **compound** are held together with chemical **bonds**. The type of bonding affects the properties of the material.

Type of bonding	Ionic	Simple molecular covalent	Giant molecular covalent
How the bonds form	metal atoms lose one or more electrons to form positive **ions** – non-metal atoms gain electrons to form negative ions	atoms share electrons to get a full outer shell – covalent bonds hold atoms together in molecules	atoms share electrons to get a full outer shell – covalent bonds hold billions of atoms together in giant structures
Examples	sodium chloride (the main salt in sea water)	water, oxygen	quartz (the main mineral in sand)
Strength of bonds	strong	strong bonds between atoms in each molecule – weak forces holding the separate molecules together	strong bonds extending across all the atoms in a structure
Melting and boiling points	high – ionic substances are solids at room temperature	low – most simple molecular covalent substances are liquids or gases at room temperature	high – giant molecular covalent substances are solids at room temperature
Solubility	many dissolve in water	some dissolve in water	insoluble in water
Do they conduct electricity?	conduct electricity when molten or in a solution – do not conduct when solid	do not conduct electricity	do not conduct electricity – except graphite, which is a conductor

B *Different types of bonding*

Properties of ionic and covalently bonded substances

Ionic bonds form when a metal reacts with a non-metal. **Covalent bonds** form between atoms of non-metal elements. Figure B summarises how the different types of bonds form, and the properties of substances with the different types of bonding. Materials can be classified according to the type of bonding holding the atoms together. Another type of bonding is found in metals.

---

1 Methane is a compound of carbon and hydrogen. It is a gas at room temperature.
a What type of bonding holds the atoms together?
b Does methane exist as separate molecules or as a giant molecular covalent substance? Explain your answer.

2 a Which temperature do you think is nearest to the melting point of sand?
  A −50°C **B** 50°C **C** 150°C **D** 1500°C
b Explain your answer.

Properties of metallic substances

The statues in Figure A are made of iron, which is a metal. Metals are good conductors of heat and electricity and are solids at room temperature. The one exception is mercury, which is a liquid at room temperature. Metals do not dissolve in water and they all conduct electricity. Metals are also **malleable** (can be hammered into shape). The atoms in metals are held together by **metallic bonds**. This gives metals different properties from other types of substance.

C This metal is being hammered to make it into a bowl shape.

---

3 A silvery solid conducts electricity. What kind of material is it likely to be?

4 A white powder does not dissolve in water. Explain why this observation does not allow you to say whether the substance has ionic or covalent bonds.

5 Describe a set of tests on physical properties that you could carry out to determine the type of bonding in a substance. 🖉

Skills spotlight

Scientists use models to help them to explain observations.

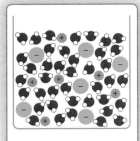

D A solution

Figure D shows a model of a dissolved ionic substance. Use this model to help you explain why ionic substances conduct electricity when they are in solution.

Learning Outcomes

4.5 Demonstrate an understanding that elements and compounds can be classified as: *a* ionic *b* simple molecular covalent *c* giant molecular covalent *d* metallic; and that each type of substance has different physical properties, including relative melting point and boiling point, relative solubility in water and ability to conduct electricity (as solids and in solution)

HSW **3** Describe how phenomena are explained using scientific models

How can we explain the properties of metals?

Many oil paints used by artists have names, for example chrome yellow or cobalt blue. These names include the name of a transition metal and it is the metal ion that helps to give the pigment its colour.

A *Colourful chemistry*

Metallic bonding

Metals are very useful materials. We use them for making cars, sculptures, buildings and tools, and even in colouring materials. The useful properties of metals include malleability and the ability to conduct electricity.

> **1** Which property of metals is useful in making:
> **a** a car body
> **b** the metal parts of an electrical plug?

All metals have only a few electrons in the outer shells of their atoms. In a solid metal, the atoms are close together and the outer electron shells overlap. The outer electrons are free to move through the structure in a 'sea of electrons'. The electrons are not located in specific atoms, so they are called **delocalised electrons**. The metal atoms form positive ions, which are held together in a regular arrangement.

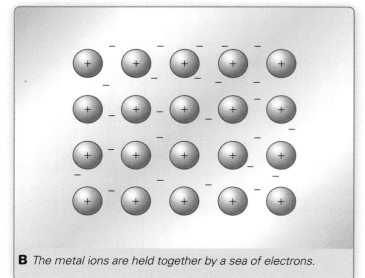

B *The metal ions are held together by a sea of electrons.*

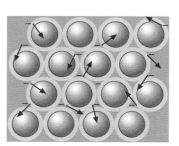
Electrons move randomly.

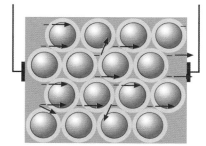
Electrons move mostly in the same direction.

C *Metals conduct electricity because the delocalised electrons can move.*

The delocalised electrons move around randomly between the positive ions in all directions. If a potential difference is applied across a piece of metal, the electrons start to drift in one direction. This movement of electrons is an electric current.

The layers of positive ions in a metal can slide over each other if a large force is applied to a piece of metal. The ions are still held together by the sea of electrons, so the metal spreads out instead of breaking. It is malleable.

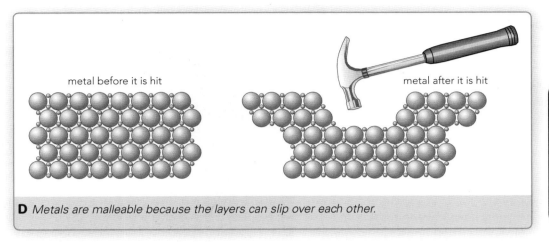

metal before it is hit   metal after it is hit

D *Metals are malleable because the layers can slip over each other.*

> **ResultsPlus**
> **Watch Out!**
>
> Transition metals form coloured compounds. Be careful not to say that they 'form colours' or that they 'are coloured'.

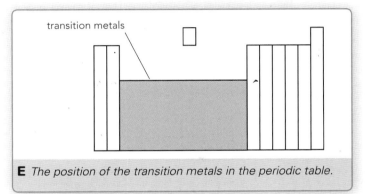

transition metals

E *The position of the transition metals in the periodic table.*

Transition metals

Most metals are **transition metals**. These are in the central block of the periodic table. Most transition metals have high melting points and form coloured compounds.

> **Skills spotlight**
>
> Many decisions about science are made for environmental reasons. Compounds of cobalt, nickel and many other transition metals are toxic. Some of these compounds are used in artists' oil paints to give them colour. Government regulations prohibit manufacturers from using these compounds in household paint. Suggest why these compounds are still allowed in artists' paints even though they are toxic to humans. Do you think this should be allowed?

2 Name four different transition metals. (You may need to look at the periodic table on page 119).

3 Write down two typical properties of transition metals.

4 Why do all metals conduct electricity?

5 Explain why metals are malleable.

6 Describe the properties of metals and how these properties are related to their uses.

Learning Outcomes

4.1 Classify elements as ... transition metals based on their position in the periodic table

4.2 Describe the structure of metals as a regular arrangement of positive ions surrounded by a sea of delocalised electrons

4.3 Describe and explain the properties of metals, limited to malleability and the ability to conduct electricity

4.4 Recall that most metals are transition metals and that their typical properties include:
  a high melting point *b* the formation of coloured compounds

HSW *13* Explain how and why decisions about uses of science and technology are made

What are the properties of the alkali metals?

Many streetlamps provide a yellowish light because they contain sodium. Sodium vapour gives out yellow light when electricity passes through it, in a similar way to its yellow flame test.

A The yellow glow from sodium lights can be seen above many cities at night.

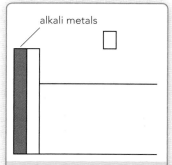

B The position of the alkali metals in the periodic table

alkali metals

The **alkali metals** are found in group 1 in the periodic table. They are solids at room temperature but they have low melting points compared to other metals (sodium melts at only 97 °C). They are all soft metals and can all be cut with a knife. The atoms in the alkali metals are held together by metallic bonding.

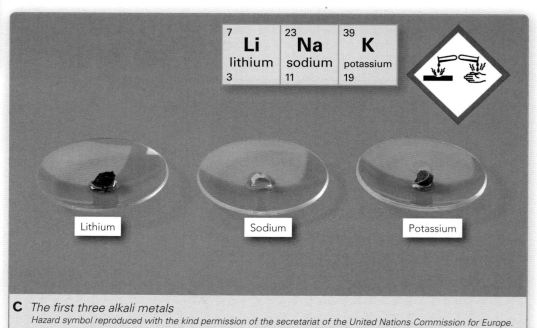

7 Li lithium 3	23 Na sodium 11	39 K potassium 19

Lithium   Sodium   Potassium

C The first three alkali metals
Hazard symbol reproduced with the kind permission of the secretariat of the United Nations Commission for Europe.

ResultsPlus
Watch Out!

Be careful that you do not say that the metals in group 1 are alkalis. They *produce* alkalis when they react with water.

1 Write down two physical properties of the alkali metals.

2 a Write a word equation to show the reaction of sodium with water.

b Write a balanced equation for the reaction.

The alkali metals are in the same group of the periodic table, so they all have similar reactions. They all react with water to form a metal hydroxide and hydrogen gas. Metal hydroxides are alkaline. For example:

lithium + water → lithium hydroxide + hydrogen

$$2Li(s) + 2H_2O(l) → 2LiOH(aq) + H_2(g)$$

The equations for the reactions of the other alkali metals are similar, because they all form ions with a charge of +1.

Reactivity

If a small piece of lithium is dropped into some water, it floats on the surface and fizzes. It gradually disappears as the reaction proceeds. Sodium reacts more vigorously than lithium and sodium also has a lower melting point. The reaction between sodium and water produces enough heat to melt the metal, which forms a molten ball that whizzes around on the surface of the water until the reaction is complete. Sometimes the hydrogen produced catches fire. Potassium reacts even more vigorously and the hydrogen catches fire, producing a lilac flame. The **reactivity** of the alkali metals increases as you go down the group.

D A ball of molten potassium reacting on water. The hydrogen produced catches fire immediately.

H

The elements at the bottom of group 1 have more electrons, and so more electron shells in their atoms, than the elements at the top of the group. This means that the outer electrons are further from the nucleus. The force between positive and negative charges is greatest when the charges are close together, so the outer electron in a caesium atom is not held as strongly as the outer electron in a lithium atom. These metals react by losing the outer electron, so this explains why caesium is more reactive than lithium.

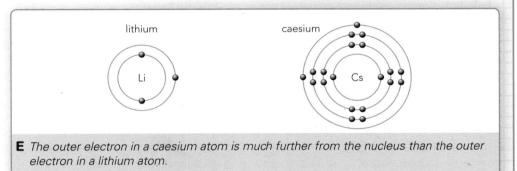

E The outer electron in a caesium atom is much further from the nucleus than the outer electron in a lithium atom.

?

3 A piece of potassium is dropped into water. Describe what you will see.

4 Universal indicator is added to a beaker of water before a piece of lithium is dropped into it. Explain what colour the indicator will be:
a before the lithium is added
b after the reaction.

H 5 Explain why potassium is more reactive than sodium.

6 Describe the differences between an alkali metal such as sodium and a transition metal such as copper.

Skills spotlight

Scientists interpret data to help them to test ideas and develop theories.
a Look at the information about the reactions of lithium, sodium and potassium with water. What trends can you see in the reactions?
b The next two elements in group 1 of the periodic table are rubidium and caesium. Predict how these two elements will react with water.

Learning Outcomes

4.1 Classify elements as alkali metals (group 1) ... based on their position in the periodic table

4.6 Describe alkali metals as: a soft metals b metals with comparatively low melting points

4.7 Describe the reactions of lithium, sodium and potassium with water to form hydroxides which are alkaline, and hydrogen gas

4.8 Describe the pattern in reactivity of the alkali metals lithium, sodium, and potassium with water and use this pattern to predict the reactivity of other alkali metals **H** and explain the pattern

HSW 2 Describe how data is used by scientists to provide evidence that increases our scientific understanding

What are the properties of the halogens?

Some types of light bulbs with tungsten filaments inside them were banned from sale in Europe in 2009 because they are not very efficient. The filament is very hot when the bulb is working and some of the tungsten evaporates. This vapour condenses and the inside of the bulb gradually becomes coated with a thin layer of metal. This reduces the efficiency of the bulb. Halogen bulbs include a small amount of bromine or iodine inside the bulb. This reacts with the tungsten vapour and prevents the outside of the bulb becoming blackened. This makes these bulbs more efficient and last longer. However, these bulbs are due to be phased out in 2016.

A *The inside of the lightbulb on the right has become coated with a thin layer of metal, that will make it less efficient. This does not happen with halogen lamps.*

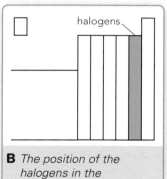

B *The position of the halogens in the periodic table*

The **halogens** are the elements in group 7, on the right-hand side of the periodic table. Fluorine and chlorine are gases at room temperature, bromine is a brown liquid and iodine is a grey solid. Fluorine gas is pale yellow and chlorine is yellow-green.

1 What colour is:
a bromine
b iodine?

2 Look at Figure C. What is the state (solid, liquid or gas) of each halogen at room temperature?

3 How do the melting and boiling points of the halogens change as you go down the group? Explain how you worked out your answer.

| 19 **F** fluorine 9 |
| 35.5 **Cl** chlorine 17 |
| 80 **Br** bromine 35 |
| 127 **I** iodine 53 |

C *The halogens are all coloured elements and they are all corrosive and toxic.*
Hazard symbols reproduced with the kind permission of the secretariat of the United Nations Commission for Europe.

Halogen reactions

The halogens all have similar reactions because they are all in the same group in the periodic table. They react with metals to form metal **halides**. For example, bromine reacts with potassium to form potassium bromide and chlorine reacts with calcium to form calcium chloride. The word 'halide' means that the compound contains only metal ions and ions of one of the halogens.

potassium + bromine → potassium bromide

$$2K(s) + Br_2(l) \rightarrow 2KBr(s)$$

calcium + chlorine → calcium chloride

$$Ca(s) + Cl_2(g) \rightarrow CaCl_2(s)$$

Each halogen needs one more electron to complete its outer electron shell, so they all form ions with a charge of –1. Fluorine is the most reactive halogen, and iodine is the least reactive. The reaction between fluorine and a metal happens much faster than reactions between metals and the other halogens.

D Chlorine reacting with iron wool

E Chlorine reacting with aluminium

6 a The formula of iron(III) chloride is $FeCl_3$. Write a word equation and a balanced symbol equation for the reaction shown in Figure D.
b Iron loses three electrons when it reacts with a halogen, and so does aluminium. Write a word equation and a balanced symbol equation for the reaction shown in Figure E.

7 Use ideas about electron structures and covalent bonding to explain why all the halogens have low melting and boiling points.

4 Name three different compounds that could be described as a 'potassium halide'.

5 Write word and balanced equations for these reactions:
a potassium + chlorine
b calcium + bromine
c sodium + fluorine

Skills spotlight

Ethical decisions are based on what people think is right or wrong. Sodium fluoride is usually present at low levels in drinking water. Some scientific studies have shown that there is less tooth decay in people who live in areas with more fluoride in the water. Fluoride is added to products like toothpaste, and in some areas it is also added to drinking water.
a Why do you think that adding fluoride to water is an ethical issue?
b Write down some of the things you need to think about before deciding whether or not fluoride should be added to drinking water.

Learning Outcomes

4.1 Classify elements as … halogens (group 7)… based on their position in the periodic table

4.9 Recall the colours and physical states of the halogens at room temperature

4.10 Describe the reactions of halogens with metals to form metal halides

HSW 13 Explain how and why decisions that raise ethical issues about uses of science and technology are made

C2.21 Displacement reactions of the halogens

How can you work out which halogens are the most reative?

The first cameras often used glass plates coated with silver halides. Many photographers used to make their own plates. Making the plates and developing them required knowing about the reactions of halogen compounds.

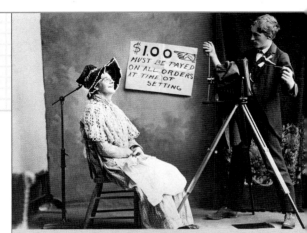

A *The exposure times for early photographs were so long that metal frames were used to make sure people kept still!*

Some elements are more reactive than others. A reactive metal will displace (take the place of) a less reactive one from a compound. For example, if you put an iron nail in a solution of copper sulfate, the nail will gradually become coated with copper. Iron is more reactive than copper and has displaced the copper in the compound. The following **displacement reaction** has taken place:

iron + copper sulfate → iron sulfate + copper

This reaction shows that iron is more reactive than copper.

chlorine water   bromine water   iodine water

B *It is easier and safer to use chlorine water (chlorine dissolved in water) than chlorine gas in this investigation. Bromine water and iodine water can also be used.*

You can use displacement reactions to work out the order of reactivity of the halogens. The word equations below show two reactions you could use to help you to work out which halogens are the most reactive:

potassium bromide + chlorine → ?

potassium chloride + bromine → ?

Your task

You are going to plan an investigation that will allow you to find out which halogens are the most reactive. Your teacher will provide you with some materials to help you organise this task.

Learning Outcomes

4.12 Investigate displacement reactions of halogens reacting with halide ions in solution

When planning an investigation like this, one of the skills you will be assessed on is your ability to identify and manage any *risks*. There are 4 marks available for this skill. Here are two extracts focusing on this skill. Other skills that you need for the assessment are dealt with in other lessons.

Student extract 1 | **A basic response for this skill**

It would be better to use the correct words for the hazards, for example, chlorine is toxic and corrosive.

> Chlorine is dangerous, so I shouldn't breathe it in.

This investigation uses chlorine water, not chlorine gas, so this comment is not relevant.

Student extract 2 | **A good response for this skill**

This student has given the correct terms for the hazards of the different substances.

> All the substances used are harmful, and chlorine water and bromine water are also toxic and corrosive. I should wear eye protection, and wash my hands straight away if I spill any of the substances on them.

Eye protection should be worn whenever corrosive substances are used.

The student has suggested more than one thing that must be done to manage the risks.

 ResultsPlus

To access 2 marks

- Identify a relevant risk which is specific to the investigation
- Suggest how to deal with that risk

To access 4 marks

- Identify most of the relevant risks which are specific to the investigation
- Describe how to manage all the risks involved

How reactive are the halogens?

Many deaths from fires in homes are caused by burning sofas and chairs. Furniture now has to be treated with 'fire retardant' compounds. These substances do not prevent fires but they give people in the house more time to escape. Many of these compounds are made using bromine.

A Both these sofas were set alight in the same way and have been burning for the same amount of time. The sofa on the left is from the UK and has been treated with a good fire retardant.

1 Write word and balanced equations for the reactions of hydrogen with:
a chlorine
b bromine.

2 Which acid is formed when hydrogen chloride dissolves in water?

Halogens and hydrogen

The halogens react with hydrogen gas to form hydrogen halides. Hydrogen halides form acids when they dissolve in water. For example, fluorine reacts with hydrogen to form hydrogen fluoride. When this is dissolved in water it forms hydrofluoric acid, HF(aq).

hydrogen + fluorine → hydrogen fluoride

$$H_2(g) + F_2(g) \rightarrow 2HF(g)$$

Skills spotlight

Hydrofluoric acid is used in etching glass. Containers of hydrofluoric acid have the hazard symbols shown in Figure C.
a What do these symbols mean?
b Why is it essential to follow the safety precautions described on the container when using hydrofluoric acid?

C Hazard symbols
Reproduced with the kind permission of the secretariat of the United Nations Commission for Europe.

B Hydrofluoric acid reacts with glass – it can be used to etch patterns in glass.

Displacement reactions

Seawater contains small amounts of bromide salts. Bromine can be displaced from a solution of bromide salts by reacting them with chlorine. For example:

sodium bromide + chlorine → sodium chloride + bromine

$$2NaBr(aq) + Cl_2(g) \rightarrow 2NaCl(aq) + Br_2(aq)$$

This is called a displacement reaction because the chlorine has 'displaced' the bromine in the salt. The solution is initially colourless but will turn an orange-brown colour as bromine is formed.

Other halogens can also take part in similar displacement reactions. For example, chlorine will displace iodide ions and the solution will turn brown-black as iodine is formed.

potassium iodide + chlorine → potassium chloride + iodine

However, if you add chlorine to a compound of fluorine, there will be no reaction. We need to look at the reactivity of all the halogens to find out why.

Reactivity of the halogens

Displacement reactions can be used to work out how reactive the different halogens are. A more reactive halogen will displace a less reactive halogen from its compounds. In the examples above, chlorine is more reactive than bromine, so chlorine displaces bromine from a bromide. Chlorine is more reactive than iodine, so it displaces iodine from an iodide.

You can carry out a series of reactions to investigate the reactivity of the halogens. Figure E shows the results of a series of reactions.

D *Chlorine water being added to potassium bromide solution (clear, colourless liquid). The orangey-yellow colour in the test tube is bromine, which has been displaced by the chlorine.*

		Salt			
		sodium fluoride	sodium chloride	sodium bromide	sodium iodide
Halogen	fluorine		reaction	reaction	reaction
	chlorine	no reaction		reaction	reaction
	bromine	no reaction	no reaction		reaction
	iodine	no reaction	no reaction	no reaction	

E *Results of halogen displacement reactions*

3 Write a balanced equation for the reaction between potassium iodide (KI) and chlorine.

4 a Which halogens will fluorine displace in a reaction?
b Which halogens will iodine displace?

5 Look carefully at Figure E and write down the halogens in order of their reactivity, starting with the most reactive element.

6 Explain how you can use a series of reactions between halogens and metal halides to work out the reactivity of the different halogen elements.

Results Plus
Watch Out!

Remember the halogens in order fluorine, chlorine, bromine, iodine then you will always know which displaces which in a reaction.

Learning Outcomes

4.11 Recall that halogens react with hydrogen to produce hydrogen halides, which dissolve in water to form acidic solutions

4.13 Describe the relative reactivity of the halogens as shown by their displacement reactions with halide ions in aqueous solution

HSW 7 Work safely, individually and with others, when collecting first-hand data

What are the properties and uses of the noble gases?

Many illuminated signs, like those in Figure A, are made using tubes filled with different noble gases. The gases glow when a high voltage is applied across the ends of the tubes.

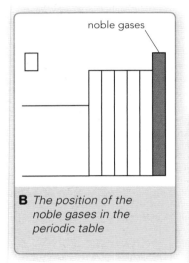

B The position of the noble gases in the periodic table

The **noble gases** are in group 0 of the periodic table. They are **inert** compared with the other elements, which means that it is very difficult to make them react with anything.

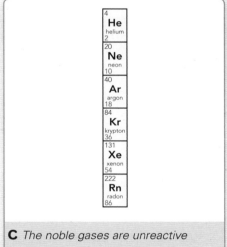

C The noble gases are unreactive

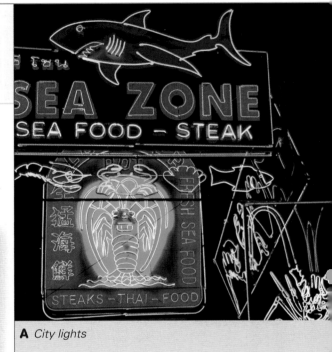

A City lights

All the elements in group 0 have full outer electron shells. Elements in other groups gain full electron shells by forming ions or by forming covalent bonds. Because noble gas atoms already have full shells, it is very difficult to make them react at all.

1 Do you think that the noble gases have high or low boiling points compared with other elements? Explain how you worked out your answer.

2 a Draw the electronic structure of a neon atom (atomic number 10).
b In what way will this drawing be similar to the electronic structures of the other noble gases?
c Use your diagram to help you to explain why the noble gases are very unreactive.

Discovery of the noble gases

Lord Rayleigh (1842–1919) was measuring the densities of different gases when he noticed that the density of pure nitrogen made in chemical reactions was less than the density of nitrogen made from the air by removing all the other gases. Sir William Ramsay (1852–1916) heard about these results, and hypothesised that the nitrogen made from air also contained a denser gas. They carried out some careful experiments to test this hypothesis and discovered a new gas, which they called argon. During this investigation, Ramsey also discovered helium. He later found neon, krypton and xenon.

Element	Density (g/cm³)
helium	0.15
neon	1.20
argon	
krypton	2.15
xenon	3.52
radon	4.40

D Densities of the noble gases

Uses of the noble gases

The noble gases are useful because they are unreactive.

- Xenon and argon were formerly used inside filament lamps, instead of air, to stop the hot filament reacting with oxygen and burning away.
- Argon and helium are used in welding, to form a blanket over the hot metal to stop it reacting with oxygen from the air.
- Argon is used in fire-extinguishing systems because it is non-flammable. Spaces such as a computer server room can be filled with argon if a fire breaks out.
- Helium has a low density and so is used for filling balloons and airships.
- Neon is often used in fluorescent lamps and advertising displays as it produces a red coloured light when electric current is passed through the tube under low pressure.

E *Welding uses an inert gas to stop the hot metal reacting.*

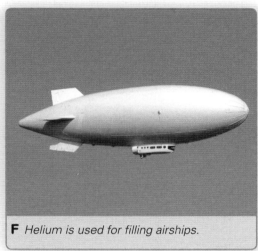

F *Helium is used for filling airships.*

3 Explain why the noble gases are non-flammable.

4 Hydrogen is the only gas with a lower density than helium. Suggest why helium is used in balloons rather than hydrogen.

5 Suggest why it was easier for early chemists to discover elements such as oxygen, rather than elements such as argon.

6 Explain why the ability to make very accurate measurements was important in the discovery of the noble gases.

Skills spotlight

Patterns in data can be used to find missing values by **interpolation**. Figure D shows the densities of some noble gases.
a Draw a graph to show the densities of the noble gases. Leave a space at the correct place for argon.
b Use your graph to estimate the density of argon.
c How could a chemist find out if this estimate is correct?

Maths skills

A line of best fit on a scatter graph can be used to estimate unknown values.

Estimating an unknown value within the range of known values is called **interpolation**. Estimating an unknown value outside the range of known values is called extrapolation.

ResultsPlus
Watch Out!

You could think of the noble gases as being like noble lords and ladies of old not having anything to do with the other common elements. They are unreactive.

Learning Outcomes

4.1 Classify elements as … noble gases (group 0) … based on their position in the periodic table

4.14 Describe the noble gases as chemically inert, compared with the other elements and demonstrate an understanding that this lack of reactivity can be explained by the electronic arrangements in their atoms

4.15 Demonstrate an understanding that the discovery of the noble gases was due to chemists: *a* noticing that the density of nitrogen made in a reaction differed from that of nitrogen obtained from air *b* developing a hypothesis about the composition of the air *c* performing experiments to test this hypothesis and show the presence of the noble gases

4.16 Relate the uses of the noble gases to their properties, including: *a* inertness … *b* low density … *c* non-flammability

4.17 Use the pattern in a physical property of the noble gases, such as boiling point or density, to estimate an unknown value for another member of the group

HSW *2* Describe how data is used by scientists to provide evidence that increases our scientific understanding

C2.24 Temperature change

How does sherbet work?

When you put sherbet in your mouth, you feel fizzing and your tongue feels cold. Sherbet contains a mixture of an acid and a base. The acid is citric acid and the base is sodium hydrogencarbonate. When they come into contact with moisture in your mouth, they react together. The reaction makes carbon dioxide, which explains the fizzing. The reaction absorbs heat energy from your mouth and so your mouth feels cold as the heat energy is taken away.

A Sherbet fizzes and feels cold on your tongue.

Many chemical and physical changes involve temperature changes. Sometimes, the mixture gets hotter and in some reactions the mixture gets colder.

Many types of chemical and physical changes are accompanied by a temperature change. Some of these include:
- dissolving – when a solid dissolves in a liquid
- neutralisation – when an acid reacts with a base
- displacement – when a metal displaces a less reactive metal from a compound
- precipitation – when two solutions are mixed and a solid is formed.

Your task

You are going to plan an investigation that will allow you to find out what happens to the temperature of a solution of copper sulfate when different masses of zinc are added and a displacement reaction takes place. Your teacher will provide you with some materials to help you organise your task.

Learning Outcomes

5.1 Measure temperature changes accompanying some of the following types of change:
a salts dissolving in water
b neutralisation reactions
c displacement reactions
d precipitation reactions

Build Better Answers

When planning an investigation like this, one of the skills you will be assessed on is your ability to record *primary evidence.* There are 4 marks available for this skill. Here are three extracts focusing on this skill. Other skills are dealt with in other lessons.

Student extract 1 — A basic response for this skill

A list like this is acceptable but it is quite hard to spot any patterns like this.

> Starting temperature 21°C. Finishing temperatures 0 = 21°C, 0.5 g = 33°C, 1.0 g = 44°C, 1.5 g = 55°C, 2.0 = 57°C, 2.5 g = 58°C.

This is good because it is a full list of results.

Don't forget units!

Student extract 2 — A better response for this skill

Including all the units and headings gives you access to higher marks.

Mass of zinc (g)	0.0	0.5	1.0	1.5	2.0	2.5
Start temperature (°C)	20	21	21	20	20	21
Final temperature (°C)	20	33	44	55	57	58

A table makes it easier to spot the patterns in the data.

Student extract 3 — A good response for this skill

Carrying out repeat experiments gives you much more data to work with.

Mass of zinc (g)	0.0	0.5	1.0	1.5	2.0	2.5
Expt 1 Start temperature (°C)	20	21	21	20	20	21
Expt 1 Final temperature (°C)	20	33	44	55	57	58
Expt 2 Start temperature (°C)	20	21	21	20	20	21
Expt 2 Final temperature (°C)	20	33	44	55	45	58

You can use your table to spot any results that do not fit the pattern. You can highlight these in your Part C work.

To access 2 marks

- Collect a suitable range of data or observations
- Record some of these appropriately

To access 4 marks

- Collect a suitable range of data or observations
- Record all your evidence appropriately
- Record repeat data or further relevant data

Why do some reactions heat things up and others cool things down?

Some foods and drinks come in self-heating cans. A chemical reaction started in one part of the can gives out lots of heat energy, heating up the food or drink in another part of the can.

A

When a chemical reaction takes place there is often an energy transfer between the reactants and the surroundings. If the energy transfer is in the form of heat energy then there is a change in temperature.

B Combustion reactions are exothermic.

Exothermic changes give out heat energy. Most chemical reactions are exothermic. The heat energy given out causes the temperature of the reaction mixture and its surroundings to increase.

All combustion reactions are exothermic reactions.

methane + oxygen → carbon dioxide + water

$$CH_4 + 2O_2 \rightarrow CO_2 + 2H_2O$$

Explosions are exothermic reactions that release a lot of heat energy and gases very quickly.

1 Give two examples each of reactions that:
a give out heat energy
b take in heat energy.

2 What is an exothermic reaction?

3 What is an endothermic reaction?

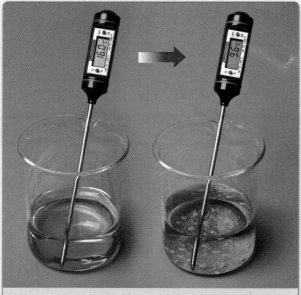

C Endothermic reactions absorb heat energy from the surroundings so the temperature decreases.

Endothermic changes take in heat energy from the surroundings, so the temperature of the surroundings decreases. Not many chemical reactions are endothermic. For example, if you react sodium hydrogencarbonate with hydrochloric acid in a test tube, the test tube feels cold because heat energy is taken from your hand.

Dissolving ammonium nitrate in water is an endothermic change. The process takes in heat energy and so the reaction mixture cools down whatever it is placed against. Photosynthesis is also an endothermic reaction.

Making and breaking bonds

In chemical reactions, bonds in the reactants are broken and new bonds are formed in the products. It takes energy to break bonds, so breaking bonds is an endothermic process. Energy is released when bonds are made, so making bonds is an exothermic process.

In an exothermic reaction, less energy is needed to break the bonds in the reactants than is released in making bonds in the products. So heat energy is released.

In an endothermic reaction, more energy is needed to break the bonds in the reactants than is released in making bonds in the products. Heat energy is absorbed.

4 a Explain why breaking bonds is an endothermic change.
b Explain why making bonds is an exothermic change.

Energy diagrams

We can show the energy changes during reactions using diagrams.

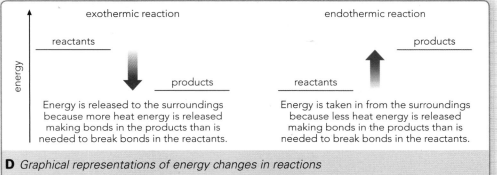

exothermic reaction

reactants

products

energy

Energy is released to the surroundings because more heat energy is released making bonds in the products than is needed to break bonds in the reactants.

endothermic reaction

products

reactants

Energy is taken in from the surroundings because less heat energy is released making bonds in the products than is needed to break bonds in the reactants.

D *Graphical representations of energy changes in reactions*

ResultsPlus
Watch Out!

Many students lose marks in exams by stating that energy is released when bonds are broken. When bonds are broken in the reactants, energy is *needed*. Remember *breakin' bonds means takin' energy.*

5 State whether each of these is exothermic or endothermic:
a copper sulfate reacts with zinc (the temperature rises) b dynamite is detonated to blast rocks c a leaf photosynthesises d coal burns e baking powder is mixed with vinegar (the temperature falls).

H 6 Draw a diagram to show the energy change for these reactions: a sodium hydrogencarbonate with hydrochloric acid b combustion of methane.

7 Explain the difference between endothermic and exothermic reactions by discussing the making and breaking of bonds.

Skills spotlight

Scientific theories can help scientists to explain and predict phenomena. In order to burn alkanes such as methane (CH_4) and ethane (C_2H_6), the bonds in these molecules must be broken. What prediction can you make about the energy needed to break all the bonds in methane compared to ethane?

Learning Outcomes

5.2 Define an exothermic change or reaction as one in which heat energy is given out

5.3 Define an endothermic change or reaction as one in which heat energy is taken in

5.4 Describe the breaking of bonds as endothermic and the making of bonds as exothermic

5.5 Demonstrate an understanding that the overall heat energy change for a reaction is: a exothermic if more heat energy is released making bonds in the products than is required to break bonds in the reactants b endothermic if less heat energy is released making bonds in the products than is required to break bonds in the reactants

H *5.6* Draw and interpret simple graphical representations of energy changes occurring in chemical reactions

HSW 3 Describe how phenomena are explained using scientific models

C2.26 Rates of reaction

What affects the rate of chemical reactions?

Rain is naturally acidic. Limestone is a rock made from calcium carbonate that weathers when it reacts with the acid in rainwater. In some areas limestone weathers faster than others. There are many factors that affect how fast limestone weathers. We can investigate several of these factors that affect how fast limestone reacts with acid.

A *These limestone rock formations in El Torcal Park, Spain, have been formed by weathering*

Chemical reactions take place at different rates. Some reactions are slow and some reactions are fast. There are several factors that affect the rate of chemical reactions, including the temperature, the surface area of the solid and the concentration of solutions.

Calcium carbonate reacts with hydrochloric acid as shown below.

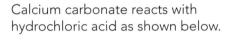

calcium carbonate (s) + hydrochloric acid (aq) → calcium chloride (aq) + carbon dioxide (g) + water (l)

$$CaCO_3(s) + 2\,HCl(aq) \rightarrow CaCl_2(aq) + CO_2(g) + H_2O(l)$$

Your task

You are going to plan an investigation that will allow you to find out how changing the concentration of the hydrochloric acid changes the rate of reaction with limestone. Your teacher will provide you with some materials to help you organise this task.

 ResultsPlus

To access 4 marks

- Provide a conclusion based on all your collected evidence
- Explain your conclusion
- Describe any relevant mathematical relationships in your conclusion

To access 6 marks

You also need to:
- Refer to the original hypothesis in your conclusion
- Refer to other scientific ideas in your conclusion

Learning Outcomes

5.7 Investigate the effect of temperature, concentration and surface area of a solid on the rate of a reaction such as hydrochloric acid and marble chips

When planning an investigation like this, one of the skills you will be assessed on is your ability to draw a *conclusion based on evidence.* There are 6 marks available for this skill. Here are three extracts focusing on this skill. Other skills are dealt with in other lessons.

Student extract 1 — A basic response for this skill

It would be better here to say that the amount of carbon dioxide produced in 30 seconds changes.

When the concentration of the acid changes, the amount of carbon dioxide changes. I can see this from my graph because there is a change in the amount of carbon dioxide which is produced when different concentrations of acid are used. There is a straight line on my graph which means that when one factor changes another factor changes by the same amount.

It is a good idea to use examples from your results or from your secondary evidence.

The student has given a mathematical relationship.

Student extract 2 — A better response for this skill

This is a clear conclusion and explains what has happened during the investigation.

Increasing the concentration of the acid makes the reaction go faster. This supports my hypothesis. I can see this from the graph of my results because there is more carbon dioxide produced in 30 seconds when the concentration of the acid is higher. The increase in the concentration of acid and in the amount of carbon dioxide produced is directly proportional.

It is a good idea to explain if your conclusion supports your hypothesis or not.

This is good use of the evidence from the practical results.

Student extract 3 — A good response for this skill

To access higher marks you should refer to scientific ideas.

You should also explain what the scientific ideas are and use them to explain what is happening in your investigation.

Increasing the concentration of the acid makes the reaction go faster. This supports my hypothesis and is what I would expect from my scientific ideas about reactions. I think that for a reaction to take place the reaction molecules need to bump into each other with enough energy to react. If you increase the number of molecules in the reaction there are more of them to bump into each other and reactions are more likely to happen. I can see this from the graph of my results because there is more carbon dioxide produced in 30 seconds when the concentration of the acid is higher. I can also see this from the graph I found in the textbook on rates of reaction. As the concentration of the acid increases, the amount of carbon dioxide produced in the time increases. The amount of carbon dioxide formed in the time is proportional to the concentration of acid.

As well as using your results you can also use secondary evidence.

What affects the speed of a chemical reaction?

Industrial chemists have to find ways of controlling the speed of chemical reactions. Some reactions need to be speeded up and some must be slowed down. Ammonia is needed to make fertilisers and chemical engineers have to find ways to speed up the slow reaction that makes ammonia.

A *Chemical engineers find ways to control the speed of chemical reactions.*

The **rate of a chemical reaction** is the speed at which it takes place. It tells us how fast products are made and reactants are used up.

Explosions are very fast chemical reactions that release a large amount of heat energy and gas in a very short time. The burning of fuels such as methane and petrol are also fast exothermic reactions. The rusting of iron and the browning of freshly peeled apples are everyday examples of slow chemical reactions.

The effect of temperature

The higher the temperature of the reactants, the faster the reaction. For example, eggs cook faster in boiling water than in warm water. Sometimes we cool reactions to slow them down. Food is put in a fridge to slow down the chemical reactions that make food go off.

B *Rusting is a slow chemical reaction between iron with oxygen and water in the air.*

The effect of concentration

Many reactions involve solutions. The **concentration** of a solution is a measure of how much solute is dissolved in a solvent. The more solute dissolved in a fixed volume of solvent, the higher the concentration of the solution. The higher the concentration of a reactant in a solution, the faster the reaction.

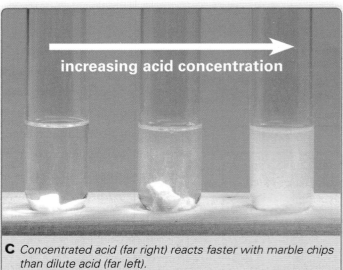

increasing acid concentration

C *Concentrated acid (far right) reacts faster with marble chips than dilute acid (far left).*

1 Give one example of a chemical reaction that at room temperature: a is very fast b is very slow.

2 Put the following chemical reactions in order of rate, from fastest to slowest: bananas ripening, oil burning, dynamite exploding, iron rusting.

3 Parts of Brazil and the UK often have the same amount of rain, but the climate is warmer in Brazil. In which of these locations will a piece of iron rust faster? Explain your answer.

The effect of surface area on the rate of a reaction

Some reactions involve solids. If a solid is broken up into smaller pieces, more of the solid is at the surface. This means that the solid has a bigger **surface area**. The greater the surface area of a solid, the faster the reaction. For example, coal is ground up into a fine powder before burning in the furnaces in a power station to help it burn faster.

Skills spotlight

Graphs can be useful for comparing sets of data. In reactions that produce gases, the volume of gas formed can be plotted against time. The gradient (steepness) of the line shows the rate.
Which of the reactions in Figure D is fastest at the start? Explain your answer.

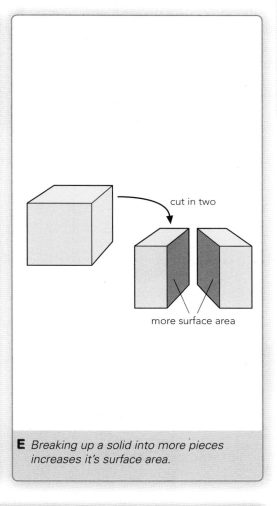

what the curve might look like if the marble was crushed into a powder

small chips

large chips

D *For small pieces of marble chip, more mass is lost as gas in the same time – the reaction is faster.*

cut in two

more surface area

E *Breaking up a solid into more pieces increases it's surface area.*

ResultsPlus
Watch Out!

Many students get confused about the link between the sizes of pieces of a solid and the surface area. Remember – when a cube is cut into two there are two new surfaces and the surface area is increased. There are more surfaces to react.

4 Explain each of the following:
a Butter goes off faster if it is left out of the fridge.
b Iron filings react faster than an iron nail with an acid.
c Concentrated cleaning solution works faster than dilute cleaning solution.

5 The reaction between a metal and hydrochloric acid is slow at room temperature. Discuss three ways this reaction could be made to go faster.

Learning Outcomes

5.8 Recall that the rates of chemical reactions vary from very fast, explosive reactions to very slow reactions

5.9 Describe the effect of changes in temperature, concentration and surface area of a solid on the rate of reaction

HSW 11 Present information using scientific conventions and symbols

How can we explain the factors that affect reaction rates?

Commercial explosives need energy to start the very fast, explosive reaction. The energy is provided by a small amount of a second explosive in a detonator inside the main explosive. The detonator is set off by a burning fuse or an electric current.

A *Energy is needed to start this explosive reaction.*

1 Give three ways of speeding up a chemical reaction.

2 What must happen for particles to react with each other?

For a reaction to occur, particles of the reacting substances must collide with each other.

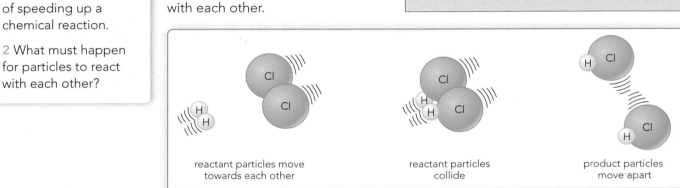

reactant particles move towards each other

reactant particles collide

product particles move apart

B *Collision and reaction of hydrogen and chlorine molecules*

The rate of any reaction will increase if the particles collide more frequently. Particles can only react with other particles when they collide, so the more often particles collide, the faster the reaction.

The collision must also have enough energy to break the bonds in the reactants. Many collisions do not have enough energy to get the reaction started. The rate of reaction will be increased if there are more high-energy collisions because the particles are more likely to react when they collide.

The effect of concentration

The more concentrated a solution, the more solute particles there are in a given volume. This means that it is more likely that reactant particles will collide with one another – and the more collisions there are between reactant particles, the faster the rate of reaction will be.

The effect of temperature

The higher the temperature, the faster the particles move. This means they collide with more energy and so are more likely to react. They also move faster at higher temperatures and so collide more frequently. This means that increasing the temperature increases the rate of reactions.

acid particles

marble chip

low concentration

acid particle meets with marble chip and reacts

marble chip

higher concentration

C *Collisions between reactant particles are more frequent at higher concentrations.*

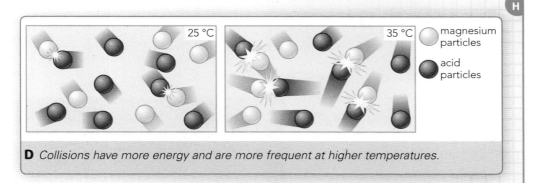

D *Collisions have more energy and are more frequent at higher temperatures.*

The effect of surface area

In a solid, only the particles on the surface can be involved in collisions. Those under the surface cannot react until those at the surface have reacted and left the solid. The greater the surface area of a solid, the more particles there are at the surface. This means that there will be more frequent collisions between reactants and so a higher rate of reaction.

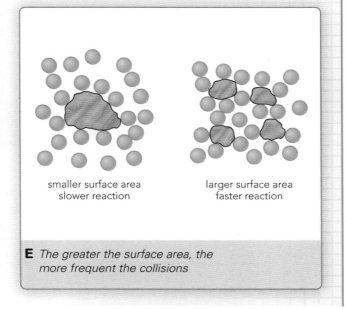

smaller surface area slower reaction

larger surface area faster reaction

E *The greater the surface area, the more frequent the collisions*

Results Plus
Watch Out!

When asked about collisions and rates of reaction, students often forget about the time of the reaction. The rate only increases if there are more collisions in one minute. The increased *frequency* of collisions increases the rate.

Skills spotlight

Models can help us to understand how complex processes happen, especially processes that we cannot see.

* Describe the main features of the collision theory model.
* Describe how these features help to explain why increasing the temperature increases the rate of a reaction.

H **3** Why do particles need energy to react when they collide?

H **4** Explain why some collisions between particles do not result in a reaction.

H **5** Concentrated bleach solutions remove stains faster than dilute solutions of bleach. Use the collision theory to explain why.

H **6** Powdered calcium carbonate reacts faster than lumps of calcium carbonate with an acid. Explain why using the collision theory.

H **7** Iron rusts faster if it is left in a warm place. Explain why using the collision theory.

H **8** Explain fully, using the collision theory, why powdered magnesium reacts faster with a warm acid than magnesium ribbon does with an acid at room temperature.

Learning Outcomes

5.10 Describe how reactions can occur when particles collide **H** and explain how rates of reaction are increased by increasing the frequency and/or energy of collisions

H **5.11** Demonstrate an understanding that not all collisions lead to a reaction, especially if particles collide with low energy

HSW **3** Describe how phenomena are explained using scientific models

29 Catalysts

What do catalysts do to reactions?

Fireworks used in stage effects must have very carefully controlled explosions otherwise they would be very dangerous. These fireworks contain many substances – some are added to control the rate of the reactions.

A

Catalysts are substances that speed up chemical reactions without being used up in the reactions. Many chemical processes use catalysts to increase the rate of production of products.

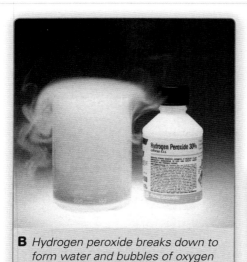

B Hydrogen peroxide breaks down to form water and bubbles of oxygen gas much faster if the catalyst manganese dioxide is added.

C

Name of catalyst	Process
iron	making ammonia from nitrogen and hydrogen
nickel	making margarine from liquid vegetable oils
aluminium oxide	for cracking alkanes

?

1 What is a catalyst?

2 Why are catalysts important in industrial processes?

3 Suggest why catalysts save money in industrial processes.

Using a catalyst often means that reactions can be done at lower temperatures and pressures than they would otherwise. This means that less energy is used, which saves money and energy resources.

Catalytic converters

The combustion of petrol in car engines produces toxic carbon monoxide and unburned hydrocarbons. Since 1993, new petrol-engine cars in the EU have been built with **catalytic converters** to reduce these pollutants in exhaust gases. Catalytic converters help to combine carbon monoxide and unburned petrol with oxygen from the air to form carbon dioxide and water vapour.

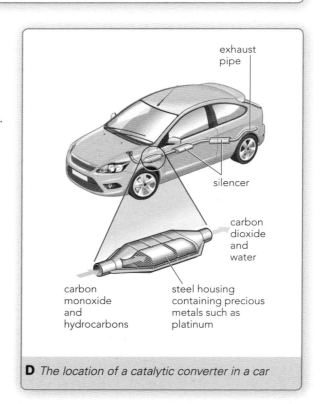

exhaust pipe

silencer

carbon dioxide and water

carbon monoxide and hydrocarbons

steel housing containing precious metals such as platinum

D The location of a catalytic converter in a car

Catalytic converters contain the transition metals platinum, rhodium or palladium. These all speed up reactions that remove the pollutants in exhaust gases. The metals used as catalysts are very expensive. To make a catalytic converter as efficient as possible while using as little of the metals as possible, scientists have refined the designs of converters to give the catalyst a high surface area. The converter is about the size of a shoe box but has a total internal surface area of about three football pitches due to a honeycomb structure. A very thin surface coat of the catalyst is used on this structure to minimise the amount of catalyst used.

Reactions are faster at higher temperatures, and catalytic converters work best at high temperatures. When an engine is first started, a catalytic converter is cool and does not work as well. However, the hot gases from the engine quickly heat it up.

E *Inside a catalytic converter*

Skills spotlight

There are many factors that affect how laws are made about the use of science and technology. Describe some of the social, economic and environmental factors in the decision to pass the law that all new cars should be fitted with catalytic converters.

ResultsPlus
Watch Out!

Students sometimes make mistakes when writing catalysts in a word equation. The catalyst must *never* be written with the reactants or the products. It should be written above the arrow.

4 a Why are cars fitted with catalytic converters?
b Which metals are used as catalysts in catalytic converters?
c What pollutants do they remove?
d What do they convert these pollutants into?

5 a Explain why catalysts are usually made with a high surface area.
b **H** Explain why a high surface area has the effect it has.
c How is a high surface area achieved in a catalytic converter?
d Why do catalytic converters work best at high temperature?

6 Suggest two reasons why we should not throw away old catalytic converters but recycle the metal catalysts.

7 A company wants to speed up the reaction it uses to make an important product. It plans to either increase the temperature or use a catalyst. Explain how each method works and suggest why using a catalyst may be the cheaper way to do it.

Learning Outcomes

5.12 Recall the effect of a catalyst on the rate of reaction

5.13 Demonstrate an understanding that catalytic converters in cars: a have a high surface area to increase the rate of reaction of carbon monoxide and unburnt fuel from exhaust gases with oxygen from the air to produce carbon dioxide and water
b work best at high temperatures

HSW 13 Explain how and why decisions about uses of science and technology are made

How do you calculate the relative mass and formula for a compound?

Some metal ions produce colours when burned – this is why they can be added to fireworks. For example, calcium ions in calcium chloride give fireworks an orange colour and lithium ions in lithium carbonate give red colours. Chemists need to know the formulae for these compounds.

A

Element	Relative atomic mass
hydrogen	1
lithium	7
carbon	12
nitrogen	14
oxygen	16
magnesium	24
aluminium	27
sulfur	32
chlorine	35.5
calcium	40
iron	56

B Relative atomic masses

The **relative atomic mass**, A_r, of an element compares the mass of an atom to the mass of a ^{12}C atom. Carbon-12 atoms have a relative atomic mass of 12. ^{1}H atoms are twelve times lighter than ^{12}C atoms so they have a relative atomic mass of 1. Figure B lists some of the common relative atomic masses.

The **relative formula mass**, M_r, of a substance is the sum of the relative atomic masses of all the atoms or ions in its formula. For example, carbon dioxide has the formula CO_2 because each molecule contains one carbon atom and two oxygen atoms:

$$M_r = A_r(C) + 2 \times A_r(O)$$
$$= 12 + (2 \times 16) = 44$$

For magnesium sulfate, $MgSO_4$,

$$M_r = A_r(Mg) + A_r(S) + 4 \times A_r(O)$$
$$= 24 + 32 + (4 \times 16) = 120$$

The brackets in the formula $Ca(NO_3)_2$ show that for each calcium ion there are two nitrate ions. This means that each unit of calcium nitrate contains one calcium, two nitrogen and six oxygen atoms:

$$M_r = A_r(Ca) + 2 \times A_r(N) + 6 \times A_r(O)$$
$$= 40 + (2 \times 14) + (6 \times 16) = 164$$

1 Calculate the relative formula masses of:
a oxygen, O_2
b water, H_2O
c ammonia, NH_3

2 Work out the number of atoms of each element present in:
a copper sulfate, $CuSO_4$ **b** iron(III) hydroxide, $Fe(OH)_3$

3 Calculate the relative formula mass for:
a magnesium hydroxide, $Mg(OH)_2$ **b** aluminium nitrate, $Al(NO_3)_3$

ResultsPlus
Watch Out!

Relative formula masses are measured *relative* to ^{12}C and have *no* units.

Empirical formulae

A substance can be represented by an **empirical formula**. This shows the simplest whole number ratio of atoms or ions of each element in the substance. The empirical formula of water is H_2O because there are twice as many hydrogen atoms as oxygen in one molecule.

A sample of a compound can be analysed to find the mass of each element present in the compound. You can then calculate the empirical formula. Figure C shows how to work out an empirical formula, using calcium chloride (made by reacting 10.0 g of calcium and 17.8 g of chlorine) as an example.

Symbol for element	Ca	Cl
Mass in g	10.0	17.8
Relative atomic mass	40	35.5
Divide the mass of each element by its relative atomic mass	$\dfrac{10.0}{40} = 0.25$	$\dfrac{17.8}{35.5} = 0.5$
Divide the answers by the smallest number to find the simplest ratio	$\dfrac{0.25}{0.25} = 1$	$\dfrac{0.5}{0.25} = 2$
Empirical formula	$CaCl_2$	

C How to calculate an empirical formula. If there are three elements in the compound add an extra column for the additional element.

The true formula for a simple molecular compound is called the **molecular formula**. This shows the actual number of atoms of each element in a molecule. Ethene has the molecular formula C_2H_4 and propene is C_3H_6 – but both have the empirical formula CH_2.

---

4 Calculate the empirical formula of the following substances:
a water, if a sample contains 0.6 g of hydrogen and 4.8 g of oxygen
b lithium carbonate, if a sample contains 2.8 g of lithium, 2.4 g of carbon and 9.6 g of oxygen
c potassium nitrate, if a sample contains 5.85 g of potassium, 2.1 g of nitrogen and 7.2 g of oxygen.

5 Write a short paragraph to show how the following are connected:
• relative atomic mass
• relative formula mass
• empirical formula
• molecular formula.

ResultsPlus
Watch Out!

Some students round the simplest ratios too much. For example, a ratio of 1 : 1.5 should be multiplied by 2 to give 2 : 3, not rounded to 1 : 2.

Skills spotlight

Scientific information can be presented concisely using symbols and formulae. Explain to a friend what is meant by the formula $Fe(OH)_2$ and how to calculate the relative formula mass.

Maths skills

Finding the empirical formula of a compound from its molecular formula is the same as **simplifying ratios**.

For example:
The molecular formula of glucose is $C_6H_{12}O_6$, which means the ratio of carbon to hydrogen to oxygen atoms is 6 : 12 : 6. To simplify this ratio we divide by a common multiple, in this case 6, to get 1 : 2 : 1. The empirical formula of glucose is therefore CH_2O.

Learning Outcomes

6.1 Calculate relative formula mass given relative atomic masses

6.2 Calculate the formulae of simple compounds from reacting masses and understand that these are empirical formulae

HSW 11 Present information using scientific conventions and symbols

>>>>>>>>>>>>>>>>>>>>>>>>>>>>

How can you find a formula?

C2.31 To determine the empirical formula for magnesium oxide

How do you work out the formula for magnesium oxide from an experiment?

Magnesium burns in air or oxygen with a bright white flame. This reaction is used in fireworks and also in flares to protect military aircraft from missiles. Infra red missiles are guided towards the heat from an aircraft's engines. Decoy flares contain magnesium that burns producing a temperature higher than the aircraft's engine so the missile hits the flare and not the aircraft.

A *Decoy flares used to protect aircraft from attack.*

Magnesium reacts with oxygen from the air to produce magnesium oxide.

magnesium + oxygen → magnesium oxide

If you carry out this experiment in a crucible you can find the mass of oxygen that reacts from the initial mass of magnesium and the mass of magnesium oxide produced.

B *Magnesium burning in a Bunsen flame*

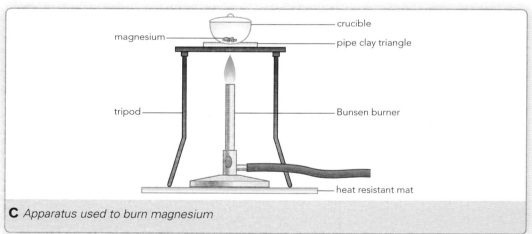

C *Apparatus used to burn magnesium*

Labels: magnesium, tripod, crucible, pipe clay triangle, Bunsen burner, heat resistant mat

Your task

You are going to plan an investigation that will allow you to find out the empirical formula for magnesium oxide. Your teacher will provide you with some materials to help you organise this task.

Learning Outcomes

6.3 Determine the empirical formula of a simple compound, such as magnesium oxide.

When planning an investigation like this, one of the skills you will be assessed on is your ability to *evaluate your method*. There are 6 marks available for this skill. Here are three extracts focusing on this skill.

Student extract 1 — A basic response for this skill

The method was difficult to do because I dropped the crucible lid. It would be easier to leave the lid off.

> You need to suggest at least one strength or weakness in the method

Although this would make the method easier it would also reduce the quality of your results because you need to keep as much magnesium oxide in the crucible as you can.

Student extract 2 — A better response for this skill

> This is a better answer because the student has identified both a strength and a weakness.

The method was quite good as all of the magnesium burnt but it was hard to lift the lid without letting the fumes out. We should have tried lifting the lid just a little before we lit the Bunsen.

> Suggesting an improvement is good but you should explain why this would improve the method and relate this back to the hypothesis.

Student extract 3 — A good response for this skill

> This is a good explanation of how the weakness in the method affects the mass of oxygen.

We measured the mass of magnesium carefully on a balance but some of the magnesium oxide escaped when we lifted the crucible lid. This made the mass of magnesium oxide lower than it should be so the mass of oxygen was lower as well. We could practice lifting the lid just a little to let less fumes out but do it more often. If we used bigger pieces of magnesium we would get more magnesium oxide so it would be easier to measure the masses so the mass of oxygen would be more accurate.

> These are good suggestions that would improve the method and give better quality data.

 Results**Plus**

To access 4 marks

- Describe strengths or weaknesses in your method
- Provide reasons for any anomalies
- Suggest how to improve your method and explain how this will improve the quality of the evidence you could collect

To access 6 marks

You also need to:
- Relate your comments back to the original hypothesis of the investigation

A *Making fireworks*

How do you find the percentage of each element in a compound?

Fireworks contain compounds called oxidising agents, which supply oxygen so that other compounds can combust. When making a firework, it is very important to know how much oxygen there is in the oxidising agent and how much oxidising agent should be added to each firework.

You can use relative masses to calculate the **percentage by mass** of each element in a compound.

Element	Relative atomic mass
hydrogen	1
carbon	12
nitrogen	14
oxygen	16
sodium	23
magnesium	24
chlorine	35.5
potassium	39
calcium	40
zinc	65

B *Some relative atomic masses*

$$\text{percentage by mass of an element in a compound} = \text{number of atoms of element} \times \frac{A_r}{M_r} \times 100$$

Figure C shows how to calculate the percentage by mass of oxygen in potassium nitrate, KNO_3, which is used as the oxidising agent in fireworks.

Relative formula mass of potassium nitrate, KNO_3	$A_r(K)$	+	$A_r(N)$	+	3	×	$A_r(O)$
	= 39	+	14	+	(3	×	16)
	= 101						
Number of atoms of oxygen in KNO_3	3						
Relative atomic mass of oxygen	16						
% of oxygen in KNO_3	3	×	$\frac{16}{101}$	×	100	=	47.5%

C *Steps to calculate percentage by mass*

(?)

1 Calculate the percentage by mass of oxygen in a molecule of:
a carbon monoxide, CO b carbon dioxide, CO_2

2 Calculate the percentage by mass of nitrogen in each of the following:
a ammonia, NH_3 b ammonium chloride, NH_4Cl c ammonium sulfate, $(NH_4)_2SO_4$

H

Calculating the masses of reactants or products

During a chemical reaction, no atoms are lost or made, they are just rearranged to make new substances. You can use relative masses and the balanced equation for a reaction to calculate the mass of a reactant or a product.

In a firework, potassium nitrate (KNO_3) is decomposed to potassium nitrite (KNO_2) and oxygen (O_2). What mass of potassium nitrate is needed in a firework to make 1.6 g of oxygen?

Write the balanced equation.	$2KNO_3 \rightarrow 2KNO_2 + O_2$
Work out the relative masses of the substances needed in the calculation. Notice that we need to multiply the relative mass of KNO_3 by 2 because there are two particles of KNO_3 in the balanced equation.	$2 \times 101 \rightarrow$ 32 $202 \rightarrow$ 32
Convert the relative masses into the units in the question. You don't really need this step, but it's useful as a reminder of the units to use.	$202\,g \rightarrow$ $32\,g$
Divide the answers by the smallest number to find the ratio.	$\dfrac{202}{32} \rightarrow \dfrac{32}{32}$ $6.3 \rightarrow$ 1
Find the mass of potassium nitrate needed to make 1.6 g of oxygen.	$6.3 \times 1.6 \rightarrow 1 \times 1.6$ $10.1 \rightarrow 1.6$
Answer	10.1 g of potassium nitrate are needed to make 1.6 g of oxygen

D *Steps to calculate a reacting mass*

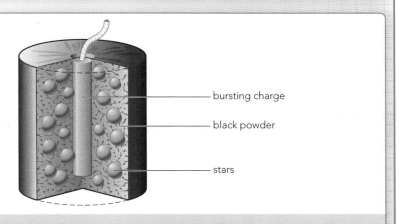

E *When this firework is blasted into the air, the fuse lights. The fuse then ignites the bursting charge and the firework explodes. Both the bursting charge and the 'black powder' contain an oxidiser (for example potassium nitrate).*

bursting charge

black powder

stars

H 3 What mass of water is formed when 4 g of hydrogen is burnt?

$2H_2 + O_2 \rightarrow 2H_2O$

H 4 What is the mass of carbon dioxide formed when 36 g of carbon is burnt?

$C + O_2 \rightarrow CO_2$

H 5 What mass of sodium chloride is formed when 10 g of sodium hydroxide reacts with hydrochloric acid?

$NaOH + HCl \rightarrow NaCl + H_2O$

Skills spotlight

Scientific ideas can be described using words and numbers. Explain, using words and relative atomic masses, how you would calculate the percentage of nitrogen in ammonium nitrate, NH_4NO_3.

6 Make notes to explain how to calculate the percentage composition by mass for a friend who has missed the lesson. Use one or more of the compounds in the following equation as an example.

$MgCO_3 + H_2SO_4 \rightarrow MgSO_4 + H_2O + CO_2$

Learning Outcomes

6.4 Calculate the percentage composition by mass of a compound from its formula and the relative atomic masses of its constituent elements

H 6.5 Use balanced equations to calculate masses of reactants and products

HSW 10 Use qualitative and quantitative approaches when presenting scientific ideas and arguments, and recording observations

How do we calculate the yield of a reaction?

Ammonia is an important chemical used to make fertilisers and nitric acid. It is made by reacting nitrogen and hydrogen, but only about 25% of the gases are converted to ammonia in the reaction.

A *The ammonia produced at Terra Nitrogen, Billingham, is used to make fertilisers.*

A factory making a certain compound needs to produce the compound as cheaply as possible. So, chemists need to make sure that the reaction converts as much of the reactants into products as possible. The amount of useful product obtained from a reaction is called the **yield**.

You can calculate the yield of reactions. For example, if you react 4 g of hydrogen with 32 g of oxygen, you expect to get 4 + 32 = 36 g of water. However, this is a **theoretical yield** and in practice you often do not get this much. The theoretical yield can be calculated from the balanced equation.

> **1** Calculate the percentage yields for these reactions:
> **a** actual yield = 2.0 g; theoretical yield = 5.0 g
> **b** actual yield = 3.2 g; theoretical yield = 4.0 g
> **c** actual yield = 19.5 g; theoretical yield = 25.0 g

ResultsPlus
Watch Out!

Students often get muddled when calculating percentage yield. Remember it is actual ÷ theoretical and the answer should always be less than 100%.

---

e.g.

Reaction between hydrogen and oxygen to form water

	$2H_2$	+	O_2	→	$2H_2O$	
Calculate formula masses	$2 \times (1+1)$	+	$(16+16)$		$2 \times (1+1+16)$	
	2×2	+	32		2×18	
Multiply according to the balanced equation	4		32		36	
Put in units	4 g		32 g	→	36 g	

So, 36 g of water should theoretically be produced when 4 g of hydrogen reacts with 32 g of oxygen.

Maths skills

To find the **percentage** of one number out of another you divide the first number by the second number and multiply by 100.

Percentage yield is the percentage of the actual yield out of the theoretical yield.

The amount of product obtained is known as the **actual yield** of the reaction. The **percentage yield** compares the actual yield to the theoretical yield.

$$\text{percentage yield} = \frac{\text{actual yield}}{\text{theoretical yield}} \times 100$$

e.g.

For example, in the reaction above between hydrogen and oxygen the actual yield was 30 g. So:

$$\text{percentage yield} = \frac{30}{36} \times 100 = 83.3\%$$

Why is the yield less than expected?

The theoretical yield of a reaction assumes that all the reactants are turned into products, and that the products are successfully separated from the reaction mixture. There are three main reasons why reactions do not give 100% yields:

- the reaction may be incomplete – this means that not all of the reactants are used up and some are left at the end
- some of the product is lost during the practical preparation – for example, when a liquid is transferred from one container to another, some of it will be left behind
- there may be other unwanted reactions taking place – for example, some of the reactants may react in different ways to make a different product.

B *Calcium oxide is produced by heating calcium carbonate in a lime kiln. Modern lime kilns are much better at converting all the calcium carbonate than older ones.*

Ⓗ **2** When limestone is heated, the calcium carbonate decomposes to form calcium oxide:

$$CaCO_3 \rightarrow CaO + CO_2$$

a Calculate the theoretical yield of calcium oxide that could be made from 125 tonnes of limestone. Use the periodic table in Appendix A to find the relative atomic masses.
b The actual yield of calcium oxide made is 45.5 tonnes. Calculate the percentage yield.

Skills spotlight

There are benefits and drawbacks with using technology. Suggest some benefits and drawbacks of increasing the pressure in a reaction between nitrogen and hydrogen gases to increase the yield of ammonia to 50%.

Ⓗ **3** One of the steps in the production of sulfuric acid is:

$$2SO_2 + O_2 \rightarrow 2SO_3$$

a Calculate the theoretical yield of sulfur trioxide that could be obtained from 256 tonnes of sulfur dioxide.
b The actual yield of sulfur trioxide produced is 202 tonnes. Calculate the percentage yield.

Ⓗ **4** Iron is extracted from iron oxide by reducing with carbon:

$$Fe_2O_3 + 3CO \rightarrow 2Fe + 3CO_2$$

a Calculate the theoretical yield of iron that could be obtained from 320 tonnes of iron oxide.
b The actual yield of iron produced is 89.6 tonnes. Calculate the percentage yield.

5 Explain the reasons why the actual yield of a reaction may be less than the theoretical yield. ✎

Learning ⊙utcomes

6.6 Recall that the yield of a reaction is the mass of product obtained in the reaction

6.7 Demonstrate an understanding that the actual yield of a reaction is usually less than the yield calculated using the chemical equation (theoretical yield)

6.8 Calculate the percentage yield of a reaction from the actual yield and the theoretical yield

6.9 Demonstrate an understanding of the reasons why reactions do not give the theoretical yield due to factors, including:
  a incomplete reactions b practical losses due to the preparation c competing, unwanted reactions

HSW 12 Describe the benefits, drawbacks and risks of using new scientific and technological developments

How does the chemical industry minimise waste and make a profit?

Aluminium is extracted from its ore in a smelter. This process creates solid and gaseous fluoride waste, which destroyed the farmland near to this smelter. In older designs in the 1940s, 12 kg of fluoride waste was produced for each tonne of aluminium. Chemists have worked to reduce this waste to about 0.5 kg per tonne and the farmland has recovered.

A An aluminium smelter

> **1** Name three products made in the chemical industry.

Disposal of waste products

The chemical industry manufactures many useful substances such as pesticides, fertilisers, cement, metals, plastics and medicines.

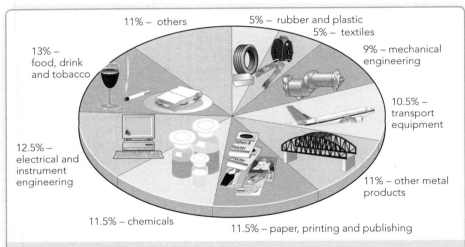

11% – others
5% – rubber and plastic
5% – textiles
13% – food, drink and tobacco
9% – mechanical engineering
10.5% – transport equipment
12.5% – electrical and instrument engineering
11% – other metal products
11.5% – chemicals
11.5% – paper, printing and publishing

B Some products formed in the chemical industry

C Metal ions in factory waste – such as copper, chromium, cadmium or mercury – may be toxic and damage the environment.

> **2** Ethanol can be produced by the fermentation of glucose:
>
> glucose → ethanol + carbon dioxide
>
> What is the by-product in this reaction?
>
> **3** Suggest why it is expensive to dispose of waste chemicals.

However, many chemical reactions produce substances other than the substance that is wanted. The additional substances are called **by-products**. For example, when chlorine is produced during the electrolysis of sodium chloride solution, sodium hydroxide and hydrogen are also produced and are by-products.

Some by-products, such as sodium hydroxide and hydrogen, are useful and can be sold, but many are useless. These useless products are called **waste products** and have to be disposed of. Today, strict laws apply to the disposal of substances that could cause pollution. Disposal can:

- be expensive – the waste may have to be transported to a landfill site or it may have to be treated with another substance to make it safe
- cause environmental problems
- cause social problems, for example, house prices could drop if a chemical plant is built near to them or there may be unpleasant smells emitted.

Finding the most cost effective process

In the UK, the law says that all chemical factories must consider the environmental impact of their production methods, including the disposal of any waste. Chemical companies also want to use reactions that will make the most profit. To do this, they try to use reactions in which:

- the percentage yield is high – low yields increase costs
- all the products of the reaction are useful so there are no waste products
- the reaction takes place quickly – the longer it takes the more expensive it becomes.

For example, when ammonia is made there are no waste products:

$$N_2 + 3H_2 \rightarrow 2NH_3$$

When iron is extracted from iron oxide there is always waste carbon dioxide:

$$Fe_2O_3 + 3CO \rightarrow 2Fe + 3CO_2$$

However, if a use can be found for waste products then a reaction becomes more economically viable.

D Iron ore contains many substances apart from iron oxide. During iron extraction, these substances form a waste called slag. This used to be piled up in to huge hills. Today, it is used to make roads.

H 4 Explain, in terms of atoms, why there is no waste product when ammonia is made.

H 5 Suggest some possible uses for the carbon dioxide produced when iron is extracted from iron oxide.

H 6 State four ways of increasing the rate of a reaction.

7 Using ideas about waste products, suggest why people object when there is a proposal to build a new chemical plant to manufacture plastics near to their homes.

ResultsPlus
Watch Out!

Some students think that all waste products are harmful – but even water can be a waste product!

Learning Outcomes

6.10 Demonstrate an understanding that many reactions produce waste products which: **a** are not commercially useful **b** can present economic, environmental and social problems for disposal

H 6.11 demonstrate an understanding that chemists in industry work to find the economically most favourable reactions where: **a** the percentage yield is high **b** all the products of the reaction are commercially useful **c** the reaction occurs at a suitable speed

HSW 13 Describe the social, economic and environmental effects of decisions about the uses of science and technology

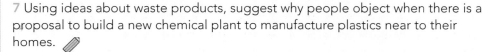

These questions are indicative of the type of questions us *in the exam. Refer to page 6 for information on the grade*

Elements in the periodic table

1. The elements chlorine, bromine and iodine are found in one group of the periodic table.

    (a) (i) Draw a straight line from each photograph to the name of the element it shows.

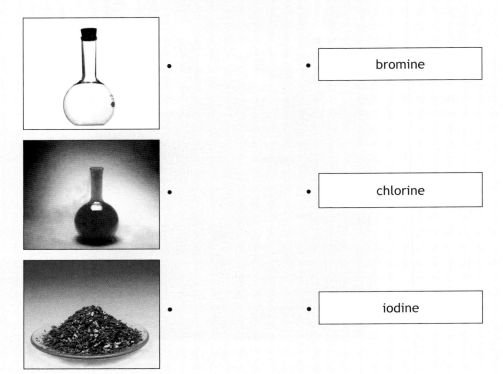

bromine	
chlorine	
iodine	

(2)

    (ii) Use the periodic table to identify the group in which these elements are found. (1)

    (iii) Name another element found in the same group as these elements. (1)

    (b) (i) Potassium is in group 1 of the periodic table. Potassium reacts with water to form potassium hydroxide and hydrogen. Write the word equation for this reaction. (1)

    (ii) When red litmus paper is dipped into potassium hydroxide solution, it turns blue. Explain why. (1)

    (iii) Potassium is a metal. The structure of potassium is

    A potassium atoms joined by covalent bonds
    B potassium ions joined by ionic bonds
    C potassium ions surrounded by a sea of electrons
    D potassium molecules

(1)

    (c) The noble gases are found in group 0 of the periodic table. They are very unreactive. Explain why. (2)

Elements and compounds

2. The diagram shows two chlorine atoms.

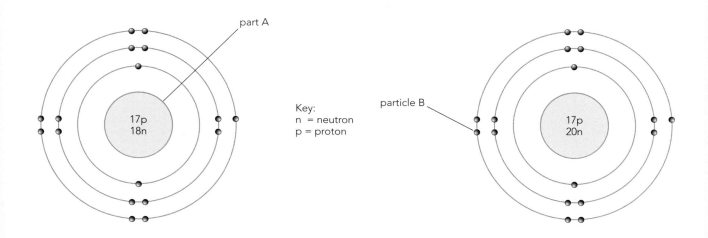

(a) Look at the diagram. What are the names of:

    (i) part A of the atom, which contains the protons and neutrons   (1) ▬

    (ii) particle B.   (1) ▬

(b) Both atoms in the diagram must be chlorine atoms because they

    A   have the same mass number
    B   have the same number of electrons
    C   have the same number of protons
    D   have three shells   (1) ▬

(c) Find magnesium, Mg, in the periodic table (see Appendix A).

    Give the group number and the period number of magnesium.   (2) ▬

(d) Magnesium reacts with oxygen, O_2, to make a compound with the formula MgO.

    (i) Give the name of the compound MgO.   (1) ▬

    (ii) Write a balanced equation for this reaction, including state symbols.   (3) ▬

    (iii) Calculate the formula mass of MgO.
        [Relative atomic masses: oxygen = 16, magnesium = 24]   (1) ▬

Marble chips and acid

3. Marble chips react with dilute hydrochloric acid. The equation for this reaction is:

$$CaCO_3 \quad + \quad 2HCl \quad \rightarrow \quad CaCl_2 \quad + \quad CO_2 \quad + \quad H_2O$$

A student carries out this reaction. The temperature of the mixture before and after the reaction is taken. The results are in the table.

Temperature at start (°C)	18
Temperature at end (°C)	47
Temperature change (°C)	

(a) (i) Complete the table to show the temperature change. (1)

(ii) Give the name of this type of reaction, where the temperature rises. (1)

(b) The reaction is carried out using the apparatus shown with one large marble chip. The mass of the apparatus and reaction mixture is measured.

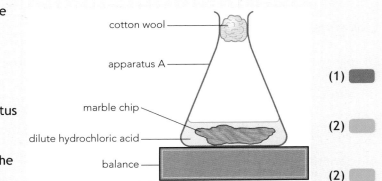

cotton wool

apparatus A

marble chip

dilute hydrochloric acid

balance

(i) Give the name of apparatus A. (1)

(ii) What would you **see** in the apparatus during the reaction? (2)

(iii) Explain why the mass reading on the balance falls during the reaction. (2)

(iv) The experiment is then repeated with an equal mass of small chips. The results are shown on the graph.

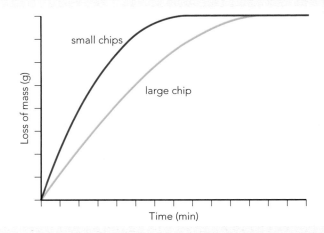

Loss of mass (g)

small chips

large chip

Time (min)

The graph shows that the size of the chips affects the rate of reaction. Explain why the rates of reaction are different. (3)

Ionic compounds

4. The diagram shows part of the structure of sodium chloride.

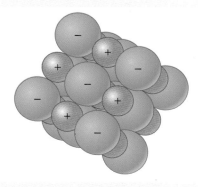

(a) Sodium chloride is ionic. It consists of sodium ions, Na^+, and chloride ions, Cl^-.
What is an ion? (1)

(b) (i) Barium sulfate is another ionic compound. It consists of barium ions, Ba^{2+}, and sulfate ions, SO_4^{2-}. The formula of barium sulfate is

   A Ba_2SO_4
   B $Ba(SO_4)_2$
   C $BaSO_4$
   D Ba_2SO_8 (1)

   (ii) Barium sulfate is used in barium meals, when X-rays are taken. Barium compounds are toxic. Explain why it is safe to drink a barium meal even though it contains a toxic compound. (2)

(c) Flame tests can be used to identify some metal ions. Explain how you could carry out a flame test to show that an unknown solid was a potassium compound. (2)

(d) Rubidium and caesium are two elements in group 1 of the periodic table.
Their compounds give distinctive flame colours. Rubidium compounds produce a red flame, but caesium compounds produce a blue flame.

Read the article below.

> In many cases, the amount of an element present in a sample is too small to see. But the element is easier to detect by spectroscopy. When the substance is placed in a flame, the hidden elements give off characteristic light. Using spectroscopy, a chemist can identify the elements by distinctive lines in the element's spectrum. In 1859, caesium was discovered in this way.

A chemist is given a sample that is thought to be a caesium compound. Evaluate the advantages and disadvantages of using a flame test or spectroscopy as methods to show whether the sample contains caesium. (6)

Structure and bonding

5. Compounds can have ionic or covalent bonding. They can be giant structures or small molecules. The diagram shows part of the structure of diamond and graphite.

diamond

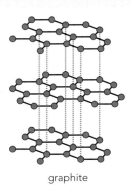
graphite

(a) (i) The same element can exist as diamond or graphite. Give the name of the element. (1)

   (ii) What type of bond joins the atoms together in diamond? (1)

(b) Water is a compound made of covalent molecules.

   (i) Complete the dot and cross diagram of a water molecule, showing outer electrons only.
   Use a cross (x) for hydrogen's electrons and a dot (●) for oxygen's electrons.

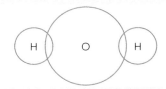

   (2)

   (ii) Calculate the percentage by mass of oxygen in water.
   [Relative atomic masses: H = 1.0, O = 16] (2)

(c) An experiment is carried out to investigate the bonding and structure in three substances.
   Their melting and boiling points are looked up in a data book. They are shaken with water
   to see if they dissolve. They are tested to see if they conduct electricity when in liquid form.

   The results are shown.

Substance	Melting point (°C)	Boiling point (°C)	Does it dissolve in water?	Does it conduct electricity when liquid?
sodium chloride	801	1413	Yes	Yes
hexane	-95	69	No	No
silicon(IV) oxide	1650	2230	No	No

   Explain, using data from the table, the bonding and structure in each of the three substances. (6)

Oxygen and the oxides of carbon

1. Oxygen gas makes up approximately 21% of the atmosphere.
   It is used in hospitals in breathing apparatus and has many uses in industry.
   It has to be separated from the other gases in the air when pure oxygen is required.

   (a) How is oxygen separated from the other gases in the air? (2)

   (b) Oxygen reacts with carbon to form either carbon monoxide, CO, or carbon dioxide, CO_2, depending on the conditions. Carbon monoxide and carbon dioxide are simple molecular, covalent compounds. Which of the following properties is shown by simple molecular, covalent compounds?

   　　A　good conductor of electricity when molten
   　　B　high boiling point
   　　C　high density
   　　D　low melting point (1)

   (c) In oxygen molecules, O_2, the atoms are held together by a double covalent bond.

   　　(i) What is a covalent bond? (1)

   　　(ii) Draw a dot and cross diagram of an oxygen molecule. (2)

   (d) 4.67 g of an oxide of carbon is found to contain 2.00 g of carbon.
   　　Show, by calculation, which oxide of carbon this is.
   　　[Relative atomic masses : C=12.0, O=16.0] (2)

Ionic compounds

2. When sodium reacts with chlorine, sodium chloride is formed.
   During this reaction ions of sodium and chlorine are formed.

   (a) Explain how ions of sodium and chlorine are formed in the reaction. (2)

   (b) Sodium ions, Na^+, and sulfate ions, SO_4^{2-}, are found in the compound sodium sulfate.

   　　(i) Which of the following is the formula for sodium sulfate?

   　　　　A　$2Na^+SO_4^{2-}$
   　　　　B　Na_2SO_4
   　　　　C　$(NaSO_4)^-$
   　　　　D　$NaSO_4$ (1)

   　　(ii) Describe how you could show the presence of sulfate ions in sodium sulfate. (3)

(c) Sodium chloride has a melting point of 801°C. Give the reason why sodium chloride has such a high melting point.

(1)

(d) A student dissolved 10g of sodium chloride in 100 cm³ of pure water.
He added lead nitrate solution and a white precipitate of lead chloride was formed.
The equation for the reaction is

$$2NaCl(aq) \quad + \quad Pb(NO_3)_2(aq) \quad \rightarrow \quad 2NaNO_3(aq) \quad + \quad PbCl_2(s)$$

(i) Calculate the maximum mass of lead chloride that would be formed from 10g of sodium chloride. Formula masses: NaCl = 58.5, $PbCl_2$ = 278]

(2)

(ii) The student filtered off the precipitate and washed it with pure water before drying it. He found the mass of the precipitate was 19.45g, which was less than the maximum mass that could be formed. Suggest a reason why the mass of precipitate formed was less than the maximum mass expected.

(1)

Atomic structure

3. The diagram shows the structure of an atom of neon.

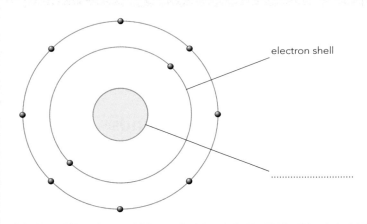

electron shell

....

(a) What is the missing label on the diagram of the neon atom?

(1)

(b) Information about the particles that make up an atom is shown below. Draw lines to connect each particle with its relative charge.

(2)

particle	relative charge
proton •	• negative
neutron •	• positive
electron •	• neutral

(c) An atom of neon has an atomic number of 10 and a mass number of 22. Using the atomic number and the mass number of neon, calculate the number of neutrons in an atom of neon.

(1)

(d) An atom of neon contains 10 electrons.

(i) Give the electronic configuration for an atom of neon.

(1)

(ii) Why should an atom of neon contain 10 electrons?

(1)

(e) Naturally-occurring neon is made up of two isotopes.
The table shows information about these isotopes.

isotope	mass number	% abundance
^{20}Ne	20	90.9
^{22}Ne	22	9.1

(i) Explain why the two isotopes have different masses.

(1)

(ii) Using the information in the table, calculate the relative atomic mass of neon.

(3)

Rates of reaction and catalytic converters

4. The rate of a chemical reaction can be changed by altering the conditions of the reaction.

(a) When magnesium metal reacts with hydrochloric acid, hydrogen gas is produced.

A piece of magnesium was added to a large excess of dilute hydrochloric acid in the apparatus shown below. The volume of gas produced was measured using a gas syringe.

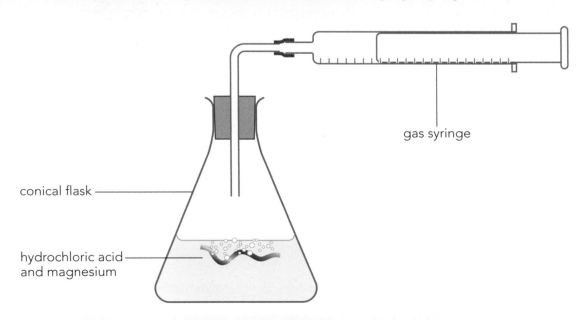

gas syringe

conical flask

hydrochloric acid and magnesium

The line marked 'A' on the graph below shows the results of the experiment.

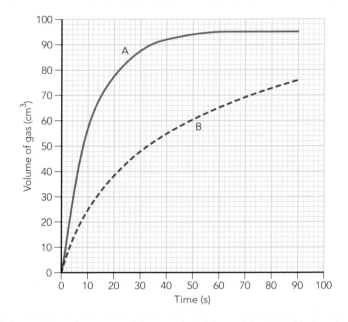

(i) The experiment was repeated using the same conditions and quantities of reactants, but a more concentrated acid was used. Copy the graph and on your copy draw a line to show the results that could have been obtained using a more concentrated acid. **(2)**

(ii) Why does the reaction stop after 95 seconds? **(1)**

(iii) The line marked 'B' shows the results of the experiment using the same conditions and quantities of reactants but using hydrochloric acid that was 10°C colder. Explain, in terms of particles, why a difference in rate of reaction occurs when there is a change in temperature. **(3)**

(b) Since 1993 all new cars in the United Kingdom have been fitted with catalytic converters.

The photograph shows a catalytic converter with a section cut away to show the inside.

Describe, with the use of appropriate chemical equations, how catalytic converters reduce the amount of pollution in exhaust gases from a car engine. **(6)**

Groups of the periodic table

5. The periodic table of elements contains groups of elements that have similar physical and chemical properties.

   (a) The transition metals form the large central block of elements in the periodic table. Which of the following properties is typical of transition metals?

       A   their compounds are all white in colour
       B   their melting points are generally low
       C   their boiling points are usually high
       D   they float when they are put into water   (1)

   (b) The halogens form group 7 of the periodic table.

      (i) Peter wants to put the halogens in order of reactivity. When he adds chlorine water to sodium bromide solution, a reaction takes place. When he adds iodine solution to sodium chloride solution, no reaction takes place. Describe what Peter has shown about the relative reactivity of the halogens and which other reaction he needs to do.   (2)

     (ii) Chlorine reacts vigorously with heated iron wool to form iron(III) chloride, $FeCl_3$. Write the balanced equation for this reaction.   (2)

   (c) The noble gases form group 0 of the periodic table. They are chemically inert compared with other elements. Why do these elements show a lack of reactivity?   (1)

   (d) The alkali metals form group 1 of the periodic table. Describe the patterns of reactivity seen within the group of alkali metals lithium, sodium and potassium and explain the patterns to justify why these three elements should all be placed in group 1 of the periodic table.   (6)

Here are three student answers to the following question. Read the answers together with the comments around and after them.

Question — Insoluble salts — Grade G–C

Lead iodide is an insoluble salt.

Describe the steps needed to prepare a pure, dry sample of lead iodide from solutions of lead nitrate and potassium iodide.

Your description should include safety precautions.

Student answer 1 — Extract typical of a level ① answer

It is correct that the substances are mixed, but be sure to name a piece of scientific apparatus for mixing. In this case a beaker could be used.

> Lead nitrate and potassium iodide are mixed in a bottle. You heat it. Then filter it. You get crystals of led iodide.

You don't need to heat the mixture in this experiment.

Take care with spelling – lead iodide appears in the question.

Summary
This answer mentions mixing the solutions and filtering, but it does not have any detail about all the steps needed. Someone should be able to follow the procedure given in the answer to make the salt. Try to think about all of the apparatus you would use. You might find it easier to draw a labelled diagram. If you cannot remember the safety precautions, ask yourself whether any of the chemicals are flammable, toxic or corrosive – in this case lead compounds are toxic so you would wear gloves.

Student answer 2 — Extract typical of a level ② answer

The answer has mentioned a safety precaution. However, there is no mention of the fact that lead compounds are toxic.

> Mix lead nitrate and potassium iodide in a beaker. Filter the mixture to get lead iodide. Put water into the filter paper to clean the salt. You should wear safety goggles to avoid getting chemicals in your eye.

A piece of apparatus is named, which is good.

Good: many answers miss out this washing stage.

Summary
You could follow this answer to make pure lead iodide, but it's a pity that the stage of drying the crystals has been forgotten. A safety precaution has been included, and it is good that the reason has been explained. However, the main point that lead compounds are toxic has been missed.

Student answer 3 | **Extract typical of a level ③ answer**

This answer makes excellent use of appropriate scientific terms and apparatus names, such as 'residue', 'beaker' and 'filter funnel'.

Lead nitrate and potassium iodide solutions are mixed in a beaker. The mixture is filtered using a filter funnel and filter paper. The residue is washed. The residue is then dried in an oven. Lead compounds are toxic, so gloves and safety glasses should be worn.

All stages have been included – mixing, filtering, washing and drying.

The answer includes the correct safety precautions for handling toxic materials.

Summary

This answer is well written. It includes all stages of the procedure to obtain a pure, dry salt, set out in the correct order. The answer uses the correct scientific terms and the correct names for pieces of scientific apparatus. This helps to make it clear and concise. A correct safety precaution is given and the reason for it is explained.

 ResultsPlus

Move from level ① to level ②

To move up from level 1 to level 2, try to include all the steps involved in the salt preparation procedure. Could someone follow your answer to carry out the experiment? Even a few names of scientific apparatus are helpful. You might find a labelled diagram a good way of communicating some of this information.

Always mention a safety precaution. Take care to write good English and spell words correctly. If a word is in the question, make sure you copy the spelling correctly.

Move from level ② to level ③

To move up from level 2 to level 3, check that you have used appropriate scientific terms in your description. Have you named the apparatus in each step?

When describing a safety precaution, give a reason for using the precaution.

Here are three student answers to the following question. Read the answers together with the comments around and after them.

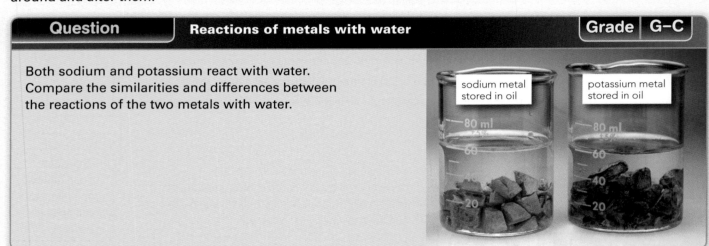

Question — **Reactions of metals with water** — **Grade | G–C**

Both sodium and potassium react with water.
Compare the similarities and differences between the reactions of the two metals with water.

sodium metal stored in oil

potassium metal stored in oil

Student answer 1 — **Extract typical of a level ① answer**

The reactions are vigorous but they do not blow up!

> When you add the metals to water they are fizzy. They wiz around. They blow up.

Try to use scientific terminology.

Summary
The answer mentions two valid observations – that the metals move around and give off a gas. These are similarities between the way the two metals react. However, the answer does not include any differences. Potassium reacts more vigorously than sodium – this should be mentioned.

Student answer 2 — **Extract typical of a level ② answer**

Good observations have been made for sodium.

It is good to mention the names of products.

> Sodium reacts with water by zooming around. It fizzes and makes a silver ball. Potassium zooms around and fizzes too. Hydrogen gas is given off. You get a flame. Potassium reacts faster than sodium.

The answer describes the way potassium reacts but does not mention the colour of the flame.

Summary
The reactions of both sodium and potassium are described, although the observations could be slightly more detailed. The answer gives the main difference between the two reactions. It might have been better to list all of the observations for sodium, and then to say how the reaction of potassium is the same (similarities) and how it is different (differences). The gas given off is identified, but not the other product. The flame colour is unusual and should be mentioned.

Sodium reacts with water by moving across the surface of the water. It fizzes and forms a silver ball. Potassium also moves across the surface and fizzes. With potassium, a purple flame is produced. Potassium reacts more violently than sodium. In both cases the reaction is:

metal + water → metal hydroxide + hydrogen

The differences between the two reactions are correctly described.

The colour of the flame is described, although strictly it is lilac rather than purple.

An equation is a good way of summarising a reaction.

Summary

Both reactions are described well. This answer is particularly good because the word equation identifies all the reactants and products. The differences in the way the two metals react are explained.

 Results**Plus**

Move from level ① to level ②

To move up from level 1 to level 2, try to give a complete description of the reactions. Imagine looking at the reaction and carefully describe everything. The question asks for the similarities and the differences, so make sure you include both. Remember that elements in group 1 have similar reactions, but get more reactive further down the group.

Move from level ② to level ③

To move up from level 2 to level 3, try to include all the reactants and products when describing the reaction. Using an equation (word or symbol) is a good way to do this. Describe all details of the reactions, including colours (lilac flame for potassium). It might also help to remember the difference between observations ('it fizzes…') and explanations ('…because hydrogen is produced').

Even the best answer above does not include the fact that the metal 'disappears'.

ResultsPlus
Build Better Answers

Here are three student answers to the following question. Read the answers together with the comments around and after them.

Question	Structures of carbon	Grade	D–A*

The diagrams show the structures of two forms of carbon, diamond and graphite.

Use the diagrams of the structures of diamond and graphite to help you explain why graphite conducts electricity but diamond does not conduct electricity.

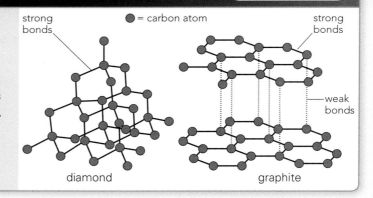

diamond   graphite

Student answer 1 — Extract typical of a level ① answer

Examiners expect answers to be written in good English. If your English is poor overall you may lose marks.

> Diamonds got lots of bonds which are strong. Graphites also got lots of strong bonds but also weak bonds and layers with spare electrons. Diamond dont have layers and no spare electrons.

Try to use the correct scientific terms. Instead of 'spare' electrons say 'delocalised' electrons. This term means that the electrons are free to move.

Summary
The answer has described what can be seen in the diagram. Also it contains the idea that there are delocalised electrons in graphite but not in diamond. However, the answer does not relate these electrons to the ability to conduct electricity.

Saying that diamond has lots of strong bonds repeats information given in the diagram, and it does not answer the question about conductivity of electricity.

Student answer 2 — Extract typical of a level ② answer

Good: the electrons are referred to as 'free', and the answer goes on to say that they are able to move. However, 'delocalised' would be even better than 'free'.

> Graphite is made up of layers. Between the layers there are free electrons. When graphite is connected to an electric current, these electrons can move. These spare electrons are not present in the diamond because there are no layers of atoms. Diamond's got lots of strong bonds between the carbon atoms, but graphite's got lots of strong bonds between atoms, but weak bonds between layers.

The explanation here for diamond is not enough to explain why diamond does not conduct electricity.

The last sentence is unnecessary. It simply repeats information from the diagram.

Summary
This answer shows a better understanding that the movement of electrons is an electric current and that it can happen in graphite. However, it does not explain why diamond does not conduct electricity.

The best answer would say that because all the outer electrons in diamond are tied up in bonds, they are not free to move and so electricity cannot flow.

Good: the answer links the number of bonding electrons in carbon to the number of bonds formed for both diamond and graphite.

In diamond, all the four electrons of carbon are involved in bonding with other carbon atoms. But in graphite, only 3 of the electrons are involved in bonding and one is 'delocalised'. These delocalised electrons can move and this is what causes the electric current to flow.

Ideally this would say 'outer-shell electrons'. Carbon has twelve electrons, but only the four outer ones are available for bonding.

The student uses the correct scientific term: delocalised.

Summary

This is a more sophisticated answer in which the number of electrons on the carbon atoms has been correctly identified with the number of bonds present in each structure. The answer also clearly states that the link between delocalised or free moving electrons with the ability to conduct an electric current.

The four outer-shell electrons of carbon are all used for bonding in diamond – hence the four strong bonds on each carbon atom, but in graphite only three are used with the fourth electron helping with the weak 'bonds' between the layers.

Another good link: the answer connects the idea of the delocalised or free electrons with the electric current.

 ResultsPlus

Move from level ① to level ②

To move up from level 1 to level 2, you need to explain the conductivity of graphite in terms of the delocalised or 'free' electrons that are between the planes of carbon atoms, which are not present in diamond.

Make sure that difficult words are spelled correctly and that the punctuation is correct.

Move from level ② to level ③

To move from level 2 to level 3, you need to consider the number of electrons in the carbon atom involved in bonding. In diamond each atom is bonded to four others, so all four outer electrons are involved in bonding – there are no delocalised electrons. In graphite each carbon atom is strongly bonded to three others, so three outer electrons are involved in bonding. The fourth electron is delocalised, which allows electric current to flow between the planes of carbon atoms.

Make sure that your descriptions use scientific language where appropriate.

Here are three student answers to the following question. Read the answers together with the comments around and after them.

Question	Compounds of chlorine	Grade	D–A*

Hydrogen chloride and sodium chloride are both compounds of chlorine.

Hydrogen chloride has a boiling point of −85 °C.
Sodium chloride has a boiling point of 1413 °C.

Explain this difference in the boiling points in terms of the structures and bonding between the particles in the two compounds.

Student answer 1   Extract typical of a level ① answer

Be sure to spell words correctly. It should be covalent.

A good description of covalent bonding.

Convalent bonding is sharing electrons. Ionic bonding is between ions.
Hydrogen chloride must have convalent bonding cos its made up of two non-metals. Sodium chloride must have ionic bonding cos its made up of a metal and a non-metal.

The question is about why the boiling points of hydrogen chloride and sodium chloride are different, but the answer doesn't mention boiling points.

Summary
The description of covalent bonding is correct, but the description of ionic bonding needs to mention the attraction between oppositely charged ions. Although the answer gives the right type of bonding for each compound, it doesn't mention the structure of each compound and how the type of bonding relates to boiling points.

Student answer 2   Extract typical of a level ② answer

Good: the answer makes the connection between weak forces and low boiling point.

Sodium chloride is made up of ions that join together in a lattice. It is not made of molecules.

Hydrogen chloride has covalent bonding between the atoms cos its a compound of two non-metals. A covalent bond is a shared pair of electrons. There are weak forces between the molecules cos the boiling point is really low. Sodium chloride must have ionic bonding in its molecules cos sodium's a metal and chlorine's a non-metal and ionic bonding gets formed between metals and non-metals. This is why it has a high boiling point.

Take care with spelling and don't use shorthand such as 'cos'.

Summary
The answer contains the correct identity of the types of bonding present in both compounds. It mentions that weak forces are linked to a low boiling point, but it says that the weak forces are because of the low boiling point rather than the other way round. It doesn't talk about the strong forces between ions and it doesn't explain the structure of sodium chloride.

A good description of an ionic structure and ionic bonding.

Sodium chloride is made up of a giant lattice of ions. Positive Na^+ ions and negative Cl^- ions attract each and are difficult to separate because they have opposite charges. Lots of energy is needed to split them up which is why it has a high boiling point. This type of bonding is called ionic bonding. Hydrogen chloride is made up of molecules, and there are weak forces between the molecules. The melting point is low because the molecules are easy to pull apart. The molecules are held together with covalent bonding, which is the sharing of a pair electrons.

A good clear description of the molecular structure.

Summary

This contains all the details needed to answer the question. It describes the structure of each compound, and how the strength of the forces between particles relates to their boiling points. The description of covalent bonding is clear, and that of ionic bonding includes a mention of opposite charges.

ResultsPlus

Move from level ① to level ②

To move up from level 1 to level 2, you need to explain how bonding and structures are linked – ionic bonding will always give rise to giant structures.

Take care when spelling difficult words, and make sure you write good English.

Move from level ② to level ③

To move up from level 2 to level 3, the answer needs to explain the observed difference of boiling points. Ionic compounds have high boiling points due to the strong forces between the ions, while simple covalent structures have low boiling points because the forces between the molecules are weak.

You should use appropriate scientific terms, justify your statements with evidence and order your arguments well.

Exam question report

Which of these is the correct balanced equation for the reaction of potassium with water?

A $K + H_2O \rightarrow KOH + H$ **B** $K + 2H_2O \rightarrow K(OH)_2 + H_2$ **C** $2K + 2H_2O \rightarrow 2KOH + H_2$ **D** $2K + 2H_2O \rightarrow K_2O + 2H_2$

Answer: The correct answer is C.

How students answered

| | 0 marks |

Most students got this wrong. They failed to spot that A has the wrong formula for hydrogen gas, B has the wrong formula for potassium hydroxide, and D has potassium oxide in it, not potassium hydroxide.

| | 1 mark |

You can actually work this out without even having to think about whether or not the equation is balanced, just by looking at the formulae!

Exam question report

An atom of an element has 24 electrons, 28 neutrons and 24 protons. What is the mass number of this atom?

Answer: 52 (the mass number is the total amount of neutrons and protons 28 + 24 = 52).

How students answered

Very few students got this correct. This is easy as long as you remember that the mass number is protons + neutrons.

ResultsPlus

Hydrogen peroxide, H_2O_2, decomposes to form water and oxygen. Describe an experiment to find the effect of temperature on the rate of decomposition of a given hydrogen peroxide solution. Use 50 cm³ samples of the hydrogen peroxide solution in the apparatus shown and describe what you would do and what reading you would take. (3)

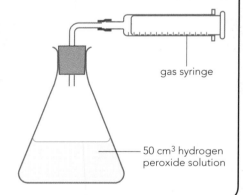

gas syringe

50 cm³ hydrogen peroxide solution

 As there are three marks, you need to make three points. A good answer would be: Heat the sample to the right temperature (1), then measure the time taken to produce a fixed volume of gas (1). Repeat the experiment several times with a new sample heated to different temperature (1).

Very few students got all three marks, and over half got no marks at all! Common mistakes were:
- Saying 'measure the time to decompose' without saying that you do this by measuring the volume of gas given off
- Describing heating the sample, but not mentioning measuring its temperature.

ResultsPlus

Sodium chloride conducts electricity when it is molten. Explain why. (1)

 Correct answer: The ions can move when it is a liquid.

Only a few students got this mark. Wrong answers included students writing about electrons moving, about atoms being free to move, about sodium being a metal, or just saying that it must be a liquid to work.

 ResultsPlus
Exam question report

Iron can be obtained from iron(III) oxide. Three mixtures were heated.
1 carbon and iron(III) oxide   2 copper and iron(III) oxide   3 magnesium and iron(III) oxide
Iron would be formed in mixture:

A 1 only   **B** 2 only   **C** 1 and 3 only   **D** 1, 2 and 3

Answer: The correct answer is C.

How students answered

Iron would be formed in mixture 1, because carbon is more reactive than iron (this is how iron is extracted from its ore). Most students answered this, and put A as their answer. Less than half of students realised that mixture 3 will also react, because magnesium is more reactive than iron.

ResultsPlus

In water molecules, atoms are covalently bonded together. Draw a dot and cross diagram of a molecule of water. (2)

 A correct answer would be something like this:

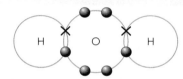

The key points are to show a pair of electrons shared between each hydrogen atom and the oxygen atom, and to show the correct total number of outer electrons in the oxygen.

Common mistakes were:
- drawing stick diagrams (H–O–H)
- trying to draw a dot and cross diagram for O–H–O or O–O–H instead of for H–O–H
- having three shells for the oxygen atom (it's ok to show just the outer shell, but oxygen only has two shells altogether!).

ResultsPlus

The carbon atoms in graphite are joined by covalent bonds. Describe the structure of graphite and explain how it is able to conduct an electric current. (1)

Correct answer: Carbon atoms are arranged in sheets (1) with each atom forming three bonds (1). There are free electrons that can move to conduct electricity (1).

Just under half of students got some marks for this, and hardly any got all three marks. Common mistakes included saying that:
- graphite has three bonds (rather than saying that all the atoms in graphite have three bonds)
- there is one free electron (rather than one per atom)
- electrons can move in the space between layers
- the layers can move over each other (this is true, but has nothing to do with why graphite conducts electricity).

Physics 2
Physics for your future

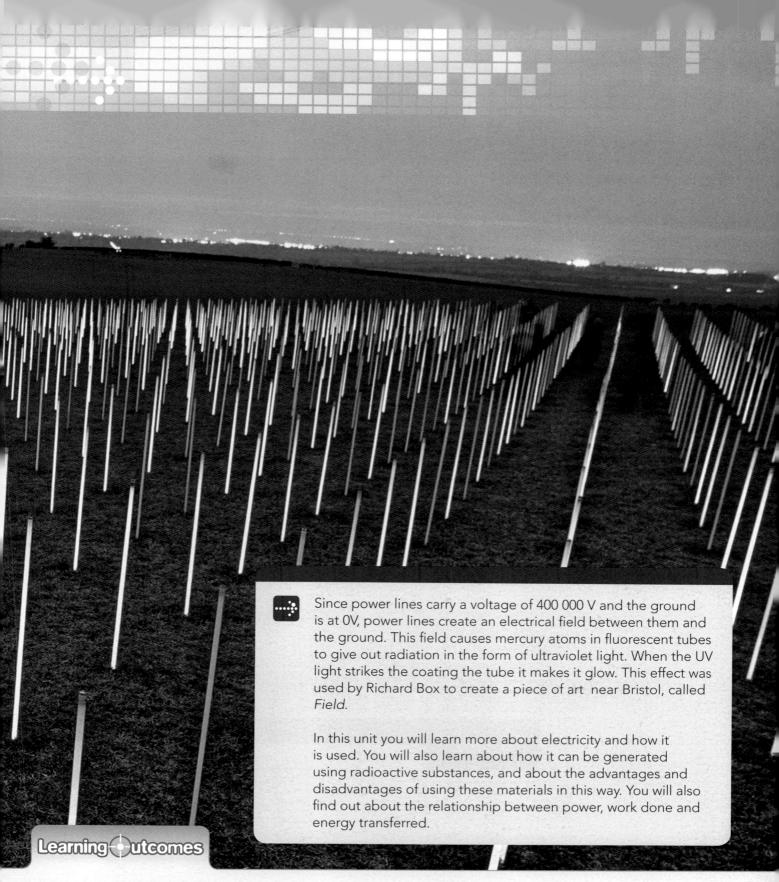

Since power lines carry a voltage of 400 000 V and the ground is at 0V, power lines create an electrical field between them and the ground. This field causes mercury atoms in fluorescent tubes to give out radiation in the form of ultraviolet light. When the UV light strikes the coating the tube it makes it glow. This effect was used by Richard Box to create a piece of art near Bristol, called *Field*.

In this unit you will learn more about electricity and how it is used. You will also learn about how it can be generated using radioactive substances, and about the advantages and disadvantages of using these materials in this way. You will also find out about the relationship between power, work done and energy transferred.

Learning Outcomes

Throughout the unit you will be required to:

0.1 Use equations given in this unit, or in a given alternate form

H *0.2* Use and rearrange equations given in this unit

0.3 Demonstrate an understanding of which units are required in equations

What causes static electricity and what can it do?

The machine in the photo is producing static electricity. The sparks it creates are similar to lightning. The man in the cage is safe because the electricity is conducted around him in the metal wires.

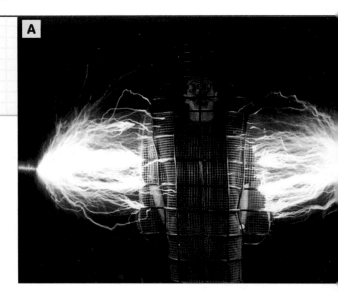

A

If you have ever had a shock when getting out of a car or heard clothing crackle as you take it off, you have experienced an effect of **static electricity**. This is caused by an electrical charge (often called an **electrostatic charge**) building up on insulating materials – including you.

All atoms contain electrically charged particles called **protons** and **electrons**. Protons have a positive charge and are found in the **nucleus** of atoms. Electrons move around the nucleus of the atom. An atom normally has no overall charge because the positive charges on the protons are balanced by the negative charges on the electrons. The nucleus also contains **neutrons**, which have no charge.

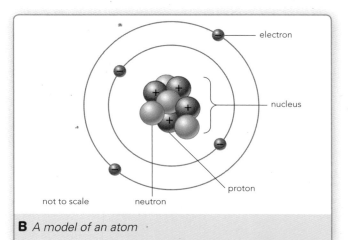

B *A model of an atom*

Skills spotlight

Scientists often use models to help them to explain why things happen, such as the model shown in Figure B. Suggest some of the differences between Figure B and a real atom.

1 What kind of materials can be given a charge?

2 Where are the following particles found in an atom:
a proton
b electron?

3 An atom has five protons. Explain how many electrons it normally has.

Property	Particle		
	proton	neutron	electron
relative charge	+1	0 (neutral)	−1
relative mass	1	1	0.0005 (negligible)

C *Particles in an atom*

If you rub two insulating materials together, electrons may be transferred from one material to the other. Protons cannot be transferred because they are fixed in the nuclei of atoms. The combination of the two materials determines which material gains electrons and which loses them.

When you rub an acetate rod with a piece of cloth, some of the electrons in the acetate move onto the cloth.

The acetate now has more protons than electrons, so it has a positive charge. The cloth has more electrons than protons, so it has an equal negative charge.

When you rub a polythene rod, some of the electrons in the cloth move onto the polythene.

D *The effect of rubbing two insulating materials together*

4 In Figure D, the bottom right drawing has no caption next to it. Write a caption for this drawing.

Objects charged with static electricity can attract or repel each other. If the charges are the same as each other (both positive or both negative) the objects will repel. If the charges on the two objects are different, they will attract each other.

You may have tried making a balloon stick to a wall by rubbing it on your jumper. Figure F explains how this works. This process is called charging by **induction** and the charge is an **induced charge**.

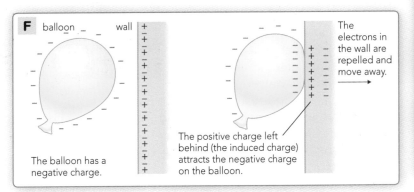

F balloon ___ wall

The balloon has a negative charge.

The positive charge left behind (the induced charge) attracts the negative charge on the balloon.

The electrons in the wall are repelled and move away.

E *This person has a charge of static electricity.*

5 Explain why only negatively charged particles can be transferred when you rub an insulating material.

6 Look at Figure E. Explain why the girl's hair is standing up.

7 You rub two balloons on your jumper. Explain why the balloons will then repel each other.

8 You rub a plastic comb with a cloth and then hold it near some small pieces of paper. Explain why the comb will pick up the pieces of paper.

ResultsPlus
Watch Out!

Students often lose marks in static electricity questions by not giving a full answer. Make sure you refer to *electrons* moving to create an overall charge, not just charged particles moving.

Learning Outcomes

1.1 Describe the structure of the atom, limited to the position, mass and charge of protons, neutrons and electrons

1.2 Explain how an insulator can be charged by friction, through the transfer of electrons

1.3 Explain how the material gaining electrons becomes negatively charged and the material losing electrons is left with an equal positive charge

1.4 Recall that like charges repel and unlike charges attract

1.5 Demonstrate an understanding of common electrostatic phenomena in terms of movement of electrons, including:
   c attraction by induction such as a charged balloon attracted to a wall and a charged comb picking up small pieces of paper

HSW **3** Describe how phenomena are explained using scientific models

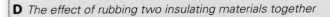

What are the uses and dangers of static electricity?

A

Helicopters build up a large charge of electricity in flight. The ground crew for this helicopter are earthing the static charge by holding up a wand that is connected to the ground with a wire. Once the helicopter has been discharged they can safely touch it.

1 You have a negative charge because your shoes have been rubbing on a carpet. Explain what happens to the charge when you touch a doorknob.

You can sometimes build up an electrostatic charge by walking on a carpet or just sitting on a chair. You only find out that you have become charged when you touch a metal object such as a doorknob. You feel a small electric shock when the electrons flow between you and the object. Electrons move from you to earth through the object or the other way around. This process is called **earthing**. You have been **discharged** (you no longer have a charge).

Charge builds up as your shoes rub on the carpet. In this case, the person is building up a negative charge by gaining electrons. The carpet is an insulating material, so the electrons cannot flow anywhere.

Electrons can flow from you, through the metal door to earth

You are discharged.

B *Earthing through a door handle*

Static electricity can build up on clouds and can cause a huge spark to form between the cloud and the ground. The lightning we see is caused by charged particles flowing through the atmosphere. The lightning also makes sound waves that we hear as thunder. Lightning strikes can kill or injure people and damage buildings.

C *Refuelling an aircraft*

The charge of static electricity that builds up on most everyday objects is usually much smaller than the charge that causes lightning. Static electricity on everyday objects is usually only dangerous when it is likely to cause a spark. If a conducting path can be provided to discharge the static electricity then there will be no spark.

Static electricity can build up as fuel flows through a refuelling pipe, and aircraft can also build up a static charge as they move through the air. A bonding line (a metal wire) is used to connect the aircraft to earth before it is refuelled. This discharges the aircraft so there will be no sparks.

A similar problem can occur when tankers deliver fuel to filling stations. In this case, the hose used to fill the underground fuel tanks is made of a conducting material so there will be no sparks.

Using static electricity

Static electricity can also be useful. It can be used in insecticide sprayers to make the spray spread out and is also used in electrostatic spray painting.

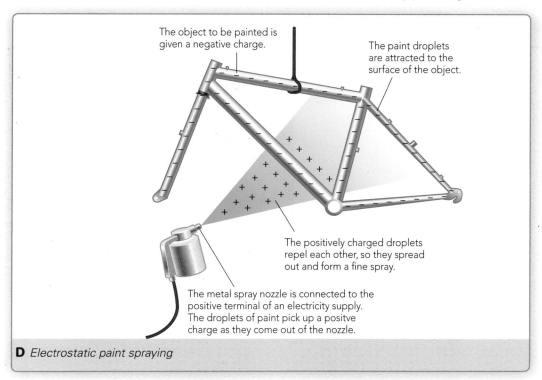

The object to be painted is given a negative charge.

The paint droplets are attracted to the surface of the object.

The positively charged droplets repel each other, so they spread out and form a fine spray.

The metal spray nozzle is connected to the positive terminal of an electricity supply. The droplets of paint pick up a positve charge as they come out of the nozzle.

D *Electrostatic paint spraying*

2 a How could a spark cause an accident when an aircraft is being refuelled?
b How does a bonding line prevent this?
c Why is it important that a bonding line is connected before the fuel nozzle touches the aircraft?

3 Look at Figure D.
a Why do the drops of paint spread out when they leave the nozzle?
b Why does the object have to be given a negative charge?

4 a Explain how using static electricity affects the spraying of insecticides.
b What are the economic effects of using static electricity?
c Suggest some possible environmental effects.

5 Describe some of the dangers of static electricity and explain how electrostatic charges can be discharged safely.

Learning Outcomes

1.5 Demonstrate an understanding of common electrostatic phenomena in terms of movement of electrons, including:
   a shocks from everyday objects b lightning

1.6 Explain how earthing removes excess charge by movement of electrons

1.7 Explain some of the uses of electrostatic charges in everyday situations, including paint and insecticide sprayers

1.8 Demonstrate an understanding of some of the dangers of electrostatic charges in everyday situations, including fuelling aircraft and tankers together with the use of earthing to prevent the build-up of charge and danger arising

ISW **12** Describe the benefits, drawbacks and risks of using new scientific and technological developments

What is an electric current?

The ULTra system is used at Heathrow Airport to carry some passengers to Terminal 5. Batteries in the vehicle supply an electric current which drives a motor. The batteries are recharged when the vehicle stops at a station.

A *The ULTra system has no driver. A computer controls all the vehicles in the system.*

An insulating material can be given a charge of static electricity because any charge transferred to the material is not conducted away.

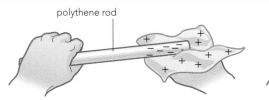

polythene rod
metal rod

Electrons are transferred from the cloth to the polythene rod. They cannot move through the polythene, so the end of the rod has a static charge.

Electrons are transferred from the cloth to the metal rod. Metals conduct electricity, so the extra electrons spread themselves out through the metal. It is difficult to detect the extra static charge.

B *Why it is difficult to charge a conducting material with static electricity*

1 What is an electric current?

2 What is the difference between conductors and insulators? Use the word 'electrons' in your answer.

All materials contain electrons. The electrons in insulating materials cannot move around but in metals some of the electrons from each atom are free to move about. If the electrons in a metal flow, there is an electric **current**.

In a piece of metal the free electrons are moving around all the time in different directions. When we put a metal wire in an electrical circuit, the cell pushes the free electrons in one direction around the circuit. There must be a complete circuit for this to happen and for the current to flow.

Cells and batteries supply **direct current**, in which the current flows only in one direction. Generators produce **alternating current**, in which the electrons change direction many times each second.

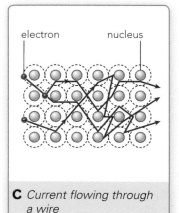

electron
nucleus

C *Current flowing through a wire*

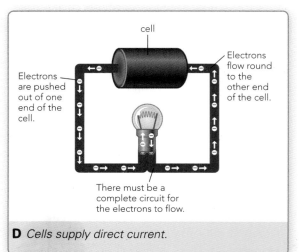

cell

Electrons are pushed out of one end of the cell.

Electrons flow round to the other end of the cell.

There must be a complete circuit for the electrons to flow.

D *Cells supply direct current.*

Charge and current

A current is a rate of flow of charged particles. The size of the current depends on how much charge is passing a point in a circuit each second. The units for charge are **coulombs** (**C**) and the units for current are **amperes** (or amps, **A**). 1 ampere is a flow of 1 coulomb of charge per second.

charge (coulombs, C) = current (amperes, A) × time (seconds, s)

This can also be written in symbols:

$$Q = I \times t$$

> **e.g.**
>
> A current of 5 A flows for 10 seconds. How much charge has flowed through the circuit?
>
> charge = current × time
>
> = 5 A × 10 s = 50 C

H This equation can be rearranged to work out the size of the current from the charge and the time.

$$\text{current (amperes, A)} = \frac{\text{charge (coulombs, C)}}{\text{time (seconds, s)}}$$

$$I = \frac{Q}{t}$$

charge

current × time

E *A formula triangle can help you to rearrange equations. Cover up the quantity you need to calculate and the symbols you can see give you the rest of the equation.*

3 a What is the difference between direct and alternating current?
b What component provides direct current?

4 What are the units for measuring: **a** charge **b** current?

5 A current of 3 A flows for 30 seconds. How much charge has flowed through the circuit?

H 6 A current of 20 A flows in a circuit. How long does it take for 5000 C of charge to flow?

7 Explain how to make a current flow in a piece of wire and why this will not work if you use an insulating material.

Learning Outcomes

1.9 Recall that an electric current is the rate of flow of charge

1.10 Recall that the current in metals is a flow of electrons

1.11 Use the equation:
charge (coulomb, C) = current (ampere, A) × time (second, s) $Q = I \times t$

1.12 Recall that cells and batteries supply direct current (d.c.)

1.13 Demonstrate an understanding that direct current (d.c.) is movement of charge in one direction only

HSW 11 Present information using scientific conventions and symbols

How do we measure current and voltage?

Electric cars usually have an ammeter mounted in the instrument panel so the driver can check how much current the engine is using. This can help them to select the best gear.

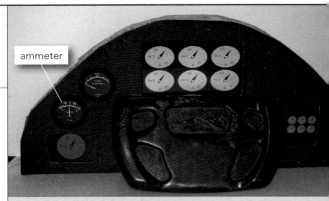

ammeter

A *An ammeter is important in solar cars. Here is a dashboard from a solar car built by a team at Glyndwr University in Wales to compete in races.*

The size of a current is measured using an **ammeter**, placed **in series** with other components in the circuit. The electrons are not 'used up' and the current leaving the cell is the same as the current that flows back into the cell, so an ammeter can be placed anywhere in a **series circuit**.

1 a Draw a series circuit with two bulbs in it.
b Draw a parallel circuit with two bulbs in it.

2 Look at Figure B. If one of the ammeters read 2 A, what would the other one read?

3 Look at Figure C. Ammeter A_2 reads 1 A and ammeter A_3 reads 3 A. What is the reading on:
a ammeter A_1
b ammeter A_4?

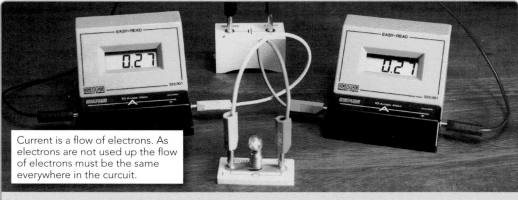

Current is a flow of electrons. As electrons are not used up the flow of electrons must be the same everywhere in the curcuit.

B *The two ammeters in this series circuit all show the same reading.*

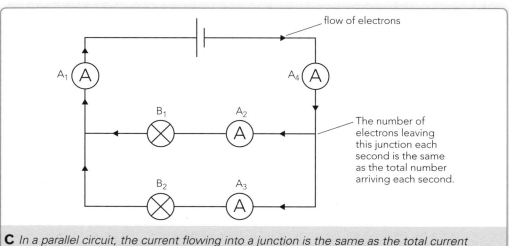

flow of electrons

The number of electrons leaving this junction each second is the same as the total number arriving each second.

C *In a parallel circuit, the current flowing into a junction is the same as the total current flowing out of it. The current is said to be 'conserved' at the junction.*

In a **parallel circuit** the current splits up when it reaches a junction. In Figure C, the reading on ammeter A_1 will be the total of the readings on ammeters A_2 and A_3.

One way to change the size of the current in a circuit is by changing the **potential difference (voltage)** of the cell or power supply. The higher the potential difference, the bigger the current.

Measuring voltage

The potential difference is a way of measuring the amount of energy transferred to a component by a current. Potential difference is measured using a **voltmeter**. The voltmeter is placed **in parallel** with the component. It measures the difference in energy between the electrons going into the component and those coming out.

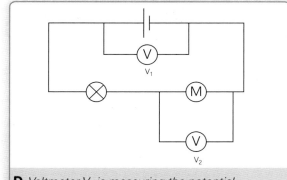

The potential difference measures the energy transferred to a component for each unit of charge that passes through that component. As energy is measured in joules and charge is measured in coulombs, 1 volt = 1 joule per coulomb. **H**

D *Voltmeter V_1 is measuring the potential difference of the cell. Voltmeter V_2 is measuring the potential difference across the motor.*

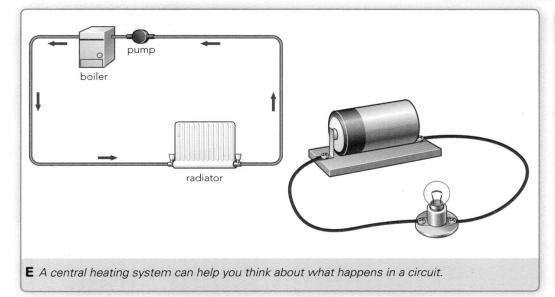

E *A central heating system can help you think about what happens in a circuit.*

Skills spotlight

Scientists often use models to help them to think about how things work. Figure E shows how a central heating system can be used as a model for a circuit. Explain which parts of the model represent each part of the circuit. What measurements in the central heating system would represent the measurements made by an ammeter and a voltmeter?

4 Look at Figure D. Draw a version of the circuit with just one voltmeter in the correct position to measure the potential difference across the bulb.

H 5 How much energy is transferred when 20 C of charge flows through a bulb if the potential difference across the bulb is 5 V?

6 Look at Figure C. Describe how you can use ammeters A_1 and A_2 to find out the current flowing through bulb B_2 and why this method will work.

ResultsPlus
Watch Out!

Potential difference is just another name for voltage. You can use either term in an examination.

Learning Outcomes

2.1 Describe how an ammeter is placed in series with a component to measure the current, in amps, in the component

2.2 Explain how current is conserved at a junction

2.3 Explain how the current in a circuit depends on the potential difference of the source

2.4 Describe how a voltmeter is placed in parallel with a component to measure the potential difference (voltage), in volts, across it

H 2.5 Demonstrate an understanding that potential difference (voltage) is the energy transferred per unit charge passed and hence that the volt is a joule per coulomb

HSW 3 Describe how phenomena are explained using scientific models

P2.5 Investigating voltage, current and resistance

What is the relationship between voltage, current and resistance?

The photo shows a fulgurite, sometimes called 'petrified lightning'. It was made when lightning struck sand. The voltage of the lightning was high enough to make a current flow for a short distance through the sand, and the heat from this current melted the sand to form this glassy structure.

A This fulgurite was found in Arizona.

The current that can flow in a circuit depends on the potential difference (voltage) and the **resistance** of the circuit. The resistance is a measure of how easy or difficult it is for electricity to flow. Resistance is measured in **ohms** (Ω).

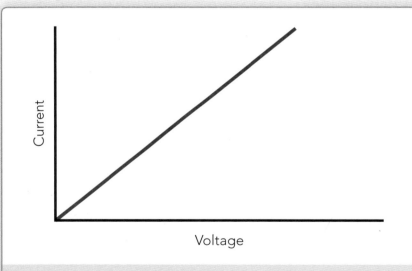

B A manufacturer would obtain this kind of graph when testing a resistor.

Resistors are used in many different types of electronic equipment to control the size of the currents. Manufacturers checking their resistors therefore need to test them at different voltages.

Your task

You are going to plan an investigation that will allow you to find out the relationship between voltage, current and resistance. You will be given resistors with different resistances. Your teacher will provide you with some materials to help you organise this task.

Learning Outcomes

2.6 Investigate the relationship between potential difference (voltage), current and resistance

Build Better Answers

When planning an investigation like this, one of the skills you will be assessed on is your ability to consider the *quality of the evidence*. There are 4 marks available for this skill. Here are two extracts focusing on this skill. Other skills are dealt with in other lessons.

Student extract 1 | A basic response for this skill

The student has correctly dealt with the anomaly (the point that did not fit the pattern) by excluding the point when drawing the graph.

> I ignored the measurement at 4 V when I drew the line on my graph for the 50 Ω resistor because it did not fit the pattern made by the other measurements. There were no measurements that seemed to be wrong in the secondary evidence.

If there are no anomalies in your evidence it is important to say so.

Commenting on the secondary evidence is also important.

Student extract 2 | A good response for this skill

The student has *explained* why it was necessary to exclude the anomalous point.

> I excluded the reading for 4 V for the 50 Ω resistor when I drew my graph because it did not seem to fit the pattern made by the other readings. The reading was probably a mistake, and if I included it when considering my conclusion, my conclusion could be wrong. I used two points from the lines of best fit when I worked out the gradients of each line, because the line of best fit evens out any errors there may have been in the ammeter or voltmeter reading.

They have shown that they have taken any possibly incorrect points into account when processing their results.

ResultsPlus

To access 2 marks

- Comment on the quality of your primary and secondary evidence
- Deal with any anomalies appropriately
- Say if you do not think there are any anomalies in your evidence

To access 4 marks

- Take account of any anomalies in your primary and secondary evidence
- Explain any adjustments you need to make to your evidence
- If you do not think there are anomalies, explain this and say that you are using all your evidence

How can the resistance of a circuit be changed?

We are used to modern cars having many automatic systems, such as headlights that come on when it gets dark, windscreen wipers that start when it rains and brakes that are applied if an obstacle is detected. Some scientists predict that automatic systems will allow driverless cars to be on the road by 2018.

A *The car is competing in a 132-mile driving competition through the desert. All the cars in the competition are driverless!*

Resistors and lamps

The size of a current flowing in a circuit depends on the potential difference of the supply and on the **resistance** of the circuit. Resistance is a way of measuring how hard it is for electricity to flow. Resistance is measured in **ohms** (Ω).

> **1** If you increase the potential difference of the power supply in a circuit, what happens to the current?

The resistance of a whole circuit depends on the resistances of the different components in the circuit. The higher the total resistance, the smaller the current.

If a resistor remains at a constant temperature, the potential difference, current and resistance are related by the equation:

potential difference (volts, V) = current (amperes, A) × resistance (ohms, Ω)

$$V = I \times R$$

> **2** A circuit has a resistance of 50 Ω and the current is 5 A. What is the voltage of the supply?
>
> **3** Resistance is not measured directly. What two measurements would you need to make to work out the resistance of a component?
>
> **H 4** A 4.5 V cell produces a current of 0.5 A in a circuit. What is the resistance of the circuit?

e.g.

What potential difference is needed to make a 2 A current flow through a 10 Ω resistor?

$$\text{potential difference} = \text{current} \times \text{resistance}$$
$$= 2\,A \times 10\,\Omega$$
$$= 20\,V$$

The current flowing in a circuit can be changed by changing the resistance. This can be done by putting a different resistor into the circuit or by using a **variable resistor**.

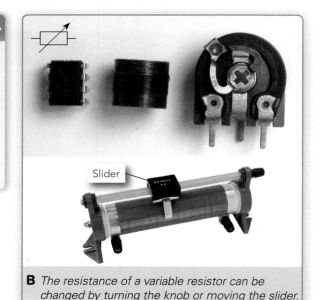

B *The resistance of a variable resistor can be changed by turning the knob or moving the slider.*

Slider

Other components

Some electrical components change their resistance depending on the potential difference or on the conditions surrounding them.

Filament lamps heat up when they are used. A higher potential difference makes them hotter, and as they get hotter the resistance increases.

Diodes conduct electricity in one direction only. If a potential difference is applied in the other direction no current will flow. The resistance of a **light-dependent resistor** (LDR) is large in the dark and gets less when light is shone on it. **Thermistors** have a high resistance when they are cold and their resistance gets less if they are heated up.

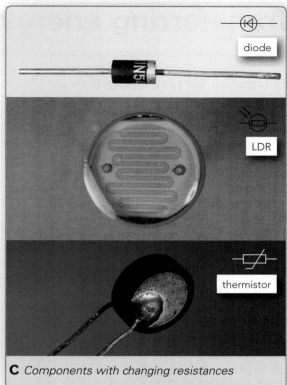

diode

LDR

thermistor

C Components with changing resistances

ResultsPlus
Watch Out!

Students sometimes get mixed up about how resistance in a circuit changes the current. Remember that for a given voltage, the current goes *down* if the resistance goes *up*.

5 You are designing an electrical system for a new car. Which of the components on this page would you need to:
a automatically switch the lights on when it gets dark
b switch the air conditioning on when it gets too hot
c adjust the volume of the radio?

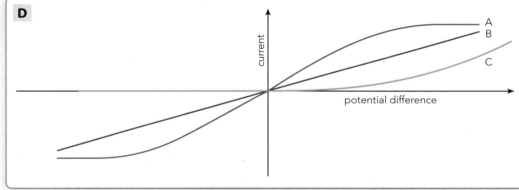

D

current

potential difference

A
B
C

6 A fan is driven by an electric motor. Explain how adding a thermistor to the circuit would make the fan move faster when the room is warmer.

Skills spotlight

Sometimes a graph can be a more effective and scientific way of describing something than using words. Figure D shows the graphs of potential difference and current for a normal resistor, a diode and a filament lamp. Which one is which?

Learning Outcomes

2.7 Explain how changing the resistance in a circuit changes the current and how this can be achieved using a variable resistor

2.8 Use the equation: potential difference (volt, V) = current (ampere, A) × resistance (ohm, Ω)
$V = I \times R$

2.9 Demonstrate an understanding of how current varies with potential difference for the following devices:
a filament lamps **b** diodes **c** fixed resistors

2.10 Demonstrate an understanding of how the resistance of a light-dependent resistor (LDR) changes with light intensity

2.11 Demonstrate an understanding of how the resistance of a thermistor changes with change of temperature (negative temperature coefficient thermistors only)

HSW 11 Present information using scientific conventions and symbols

How much energy is transferred by electric currents?

People working or playing outdoors in cold weather can wear electrically heated clothing to make sure they stay warm.

When current flows through a resistor, energy is transferred to the resistor and it becomes warm. This heating effect is useful in appliances such as electric fires and kettles. These appliances are effectively large resistors.

A similar thing happens with all components – even components such as motors where the main energy transfer is from electrical energy to kinetic (movement) energy. In the case of a motor, the heat produced is wasted energy. Some electrical appliances may produce so much waste heat energy that they become too hot to touch and could cause burns.

A An electrically heated jacket

Skills spotlight

Scientists consider the benefits, drawbacks and risks of using new scientific and technological developments. Describe what these might be when comparing an electric heater with an open log fire.

C Wires can catch fire if too much current flows. Plugs are fitted with fuses, which melt and break the circuit if the current is too high. This plug had the wrong type of fuse fitted.

1 Why do computers need cooling fans inside them? ❓

2 Give two examples of appliances where the heating effect of a current is:
a useful energy.
b wasted energy.

Ⓗ A current in a wire is a flow of electrons. As the electrons move in a metal, they collide with the **ions** in the lattice and transfer some energy to them.

Calculating power and energy

Power is the energy transferred every second. For electrical appliances the power can be worked out from the current and the potential difference. The unit of power is joules per second or **watts (W)** (1 watt = 1 joule/second).

electrical power (watts, W) = current (amperes, A) × potential difference (volts, V)

$$P = I \times V$$

e.g.

A kettle uses the mains electricity supply at 230 V. The current is 13 A. What is the power of the kettle?

power = current × potential difference

= 13 A × 230 V

= 2990 W

The total energy transferred by an appliance depends on its power and for how long it is switched on. The energy transferred can be calculated using the following equation:

energy transferred = current × potential difference × time
(joules, J)   (amperes, A)   (volts, V)   (seconds, s)

$$E = I \times V \times t$$

e.g.

The kettle in the example above takes 2 minutes to boil some water. How much energy does it transfer?

energy = current × potential difference × time

2 minutes = 120 seconds

energy = 13 A × 230 V × 120 s

= 358 800 J (or 358.8 kJ)

6 Explain why a travel kettle using a 12 V supply takes much longer to boil water than a kettle that uses the mains supply.

?

3 An electric travel blanket uses a car's 12 V supply. The current through it is 3 A.
a What is the power of the blanket?
b How much energy does it transfer if it is switched on for 10 minutes?

H 4 A soldering iron has a power of 25 W. It uses the 230 V mains supply. What is the current when it is switched on?

H 5 A travel kettle has a power of 0.5 kW. How long would it take to boil the same amount of water as in the example on the left?

Watch Out!

Students often lose marks by being careless with units. Make sure the time is converted to seconds before you put numbers into the equation. In written tests always show your working.

Learning Outcomes

2.12 Explain why, when there is an electric current in a resistor, there is an energy transfer which heats the resistor

H 2.13 Explain the energy transfer (in 2.12 above) as the result of collisions between electrons and the ions in the lattice

2.14 Distinguish between the advantages and disadvantages of the heating effect of an electric current

2.15 Use the equation:
electrical power (watt, W) = current (ampere, A) × potential difference (volt, V)
$$P = I \times V$$

2.16 Use the equation: energy transferred (joule, J) = current (ampere, A) × potential difference (volt, V) × time (second, s)
$$E = I \times V \times t$$

HSW 12 Describe the benefits, drawbacks and risks of using new scientific and technological developments

Vectors and velocity

What is velocity?

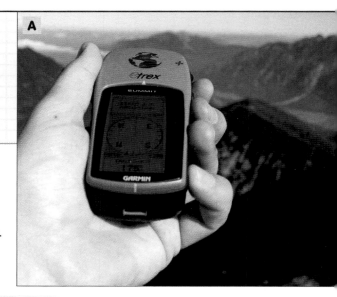

A

The Global Positioning System (GPS) uses 30 satellites in orbit around the Earth. Using the information from four of these satellites, a hand-held GPS navigation device can calculate a walker's position. It can also tell the walker their speed, in what direction they need to walk to reach their destination and how long it will take them to get there.

> **1** What is the difference between distance and displacement? ?

In a 200 m race, the runners each cover 200 m. However, the distance between the start and the finish *in a straight line* is less than this. This distance is called the **displacement**. Since displacement is measured using a straight line, it is a quantity that has a certain direction.

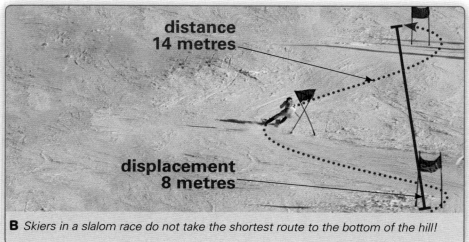

distance
14 metres

displacement
8 metres

B *Skiers in a slalom race do not take the shortest route to the bottom of the hill!*

The **speed** of an object tells you how quickly it will travel a certain distance. Speed can be calculated using the following equation:

$$\text{speed (m/s)} = \frac{\text{distance (m)}}{\text{time taken (s)}}$$

Like distance, speed only has a size (magnitude). **Velocity** tells you how quickly an object is moving but it also tells you the direction it is moving. Both speed and velocity are often measured in metres per second (m/s).

> **2** At the end of an 800 m race, a runner can have a displacement of 0 m. How is this possible? ?
>
> **3** A cyclist travels a distance of 1080 m in 2 minutes. What is their speed?
>
> **4** Velocity and force are both vector quantities. What does this mean?

Quantities like displacement and velocity, which have a size and a direction, are called **vector** quantities. Force is also a vector quantity. For example, on the Earth the weight of an object always pulls vertically downwards.

C *In this display by the Red Arrows the planes travel at the same speed, but as they travel in different directions, they have different velocities.*

Distance–time graphs

A graph in which distance is plotted against time is called a **distance–time graph**. Since time and distance are used to calculate speed, the graph can tell us various things about speed:

- horizontal lines mean the object is stationary
- straight, sloping lines mean the object is travelling at constant speed
- the steeper the line, the faster the object is travelling
- the speed can be calculated from the **gradient** of the line.

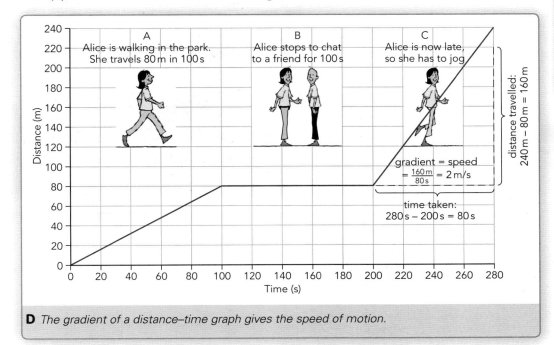

D *The gradient of a distance–time graph gives the speed of motion.*

5 A runner jogs with a velocity of 3 m/s due north. She changes direction and jogs at the same speed due west. Explain whether her velocity has changed.

H 6 The peregrine falcon is a bird that can fly at 50 m/s to catch its prey. At this speed, how far will it fly in 7 seconds?

Watch Out!

Students often lose marks when calculating speed from gradients. Look at the gradient line, find how far the object has travelled by subtracting the starting number from the finishing number. Now do the same for the time and divide the distance by the time. Remember it is the *change* in distance and time you need.

7 In Figure D, what was Alice's speed for:
a part A on the graph
b part B on the graph?

H 8 If Alice had not stopped to chat but had walked at her initial speed for 280 s, how far would she have travelled?

9 Explain how a distance–time graph can be used to give information about an object's speed.

Skills spotlight

Scientists use tables and graphs to present information. Why is showing distance–time data on a graph more useful than just showing it in a table?

Learning Outcomes

3.1 Demonstrate an understanding of the following as vector quantities:
  a displacement
  b velocity
  d force

3.2 Interpret distance / time graphs including determination of speed from the gradient

3.3 Recall that velocity is speed in a stated direction

3.4 Use the equation:
  speed (m/s) = distance (m) / time (s)

HSW **11** Present information using scientific conventions and symbols

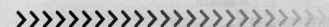

Acceleration

 ## What is acceleration?

Virgin Galactic's SpaceShipTwo can take six passengers into space. When it detaches from its carrier plane flying at a height of 15 km, a rocket accelerates the craft directly upwards to a velocity of over 2 km/s. From there, it takes just 90 seconds to reach its final height of 110 km above the Earth.

A *SpaceShipTwo is carried underneath WhiteKnightTwo.*

B *These vehicles are both accelerating in a straight line.*

Most moving things do not stay at a constant velocity. They speed up or slow down all the time. This change in velocity is called **acceleration**. Since velocity is a vector, so is acceleration – it has a size (magnitude) and a direction. If a moving object changes its velocity or direction, then it is accelerating.

Acceleration is calculated using the following equation:

$$\text{acceleration (m/s}^2) = \frac{\text{change in velocity (m/s)}}{\text{time taken (s)}}$$

Using symbols, this is written as:

$$a = \frac{(v - u)}{t}$$

where a is the acceleration
         v is the final velocity (the velocity the object ends up with)
         u is the initial velocity (the velocity the object starts with)
         t is the time taken for the change in velocity.

1 How are velocity and acceleration connected?

2 SpaceShipTwo accelerates from 0 m/s to 2100 m/s vertically upwards in 90 seconds. What is its acceleration?

e.g.

The racing car in Figure B starts from 0 m/s along the straight track and reaches a velocity of 50 m/s in 5 seconds. What is its acceleration?

$$a = \frac{(v - u)}{t} = \frac{(50\,\text{m/s} - 0\,\text{m/s})}{5\,\text{s}} = \frac{50\,\text{m/s}}{5\,\text{s}} = 10\,\text{m/s}^2$$

The acceleration tells you the change in velocity each second, so the units of acceleration are metres per second, per second. This is written as m/s² (metres per second squared). An acceleration of 10 m/s² means that each second the velocity increases by 10 m/s.

Acceleration does not always mean getting faster. An acceleration can also cause an object to get slower.

During part of the landing sequence shown in Figure C, the Space Shuttle slowed down from 70 m/s to 20 m/s in 20 seconds.

C *The Space Shuttle used its brakes once the parachute had slowed it to about 70 m/s.*

3 If a sports car accelerates at 3 m/s² from rest (0 m/s) along a straight road, what will its velocity be after 4 seconds?

ResultsPlus
Watch Out!

Students often assume that acceleration means that an object is speeding up. Take care to use the correct values for initial and final velocities, as the object may be slowing down.

$$\text{acceleration (m/s}^2) = \frac{\text{change in velocity (m/s)}}{\text{time taken (s)}} = \frac{(v - u)}{t}$$

$$a = \frac{20\,\text{m/s} - 70\,\text{m/s}}{20\,\text{s}} = \boxed{\frac{-50\,\text{m/s}}{20\,\text{s}}} = -2.5\,\text{m/s}^2$$

The minus sign shows that the Space Shuttle is slowing down. Acceleration is a vector quantity. Here the acceleration is acting in the *opposite* direction to the direction in which the vehicle is travelling.

Skills spotlight

Scientists use standard units for measurements. In the UK, car manufacturers often use the time taken for a car to accelerate from 0 mph to 60 mph to describe its acceleration. Suggest why these units are different from the units used by scientists and designers.

4 Explain why acceleration is a vector quantity.

5 A car travels along a straight road at 40 m/s. The driver brakes and brings the car to a halt in 8 seconds. What is the car's acceleration?

H 6 A train is travelling at 35 m/s. Coming into a station, it slows down with an acceleration of 0.5 m/s². How much time does it take to stop?

7 Acceleration is a vector quantity. Explain how negative, positive and zero accelerations would change the velocity of a moving object.

Learning Outcomes

3.1 Demonstrate an understanding of the following as vector quantities:
    c acceleration
3.5 Use the equation:
    acceleration (metre per second squared, m/s²) = change in velocity (metre per second, m/s) / time taken (second, s)

$$a = \frac{(v - u)}{t}$$

HSW 11 Present information using scientific conventions and symbols

How can a graph show changes in an object's velocity?

O n 15 October 1997 the *ThrustSSC*, driven by Royal Air Force pilot Andy Green, broke the land speed record. It recorded an average speed of 763.035 mph over a straight stretch of 1 mile in the Black Rock Desert – faster than the speed of sound. This is still the world record.

A

A velocity-time graph shows how the velocity of an object changes with time. Figure B shows velocity-time graphs for cars moving in different ways along a straight road.

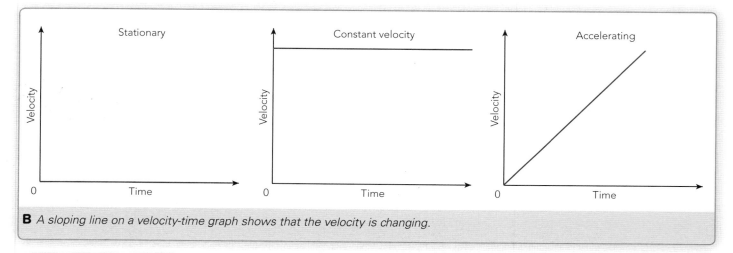

B *A sloping line on a velocity-time graph shows that the velocity is changing.*

1 What does a horizontal line on a velocity-time graph tell you about an object's velocity?

- A horizontal line means the object is travelling at constant velocity. The higher the line, the higher the velocity.
- A sloping line shows that the object is accelerating. The steeper the line, the greater the acceleration.

Figure C is a velocity-time graph for a train travelling along a straight track.

2 a In which part of Figure C is the train travelling at a constant velocity?
b In which part of the graph does the train have its greatest acceleration?
c Which part of the graph shows that the train is slowing down?

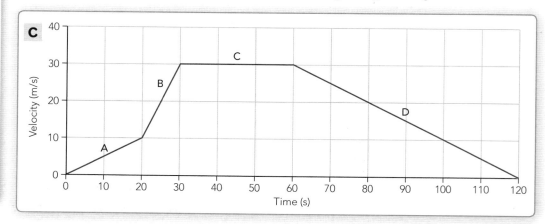

Calculations using a velocity-time graph

The acceleration of an object can be calculated from a velocity-time graph.
From the graph, find the change in velocity and the time taken for that change. Use these to calculate the acceleration.

For the graph in Figure D: $\text{acceleration} = \dfrac{8\,\text{m/s}}{4\,\text{s}} = 2\,\text{m/s}^2$

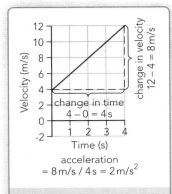

acceleration
= 8 m/s / 4 s = 2 m/s²

D *Calculating acceleration from a velocity-time graph.*

Calculating distance

The area under a velocity–time graph tells us the distance the object has travelled. The distance travelled in the first 5 s is the area of a rectangle. The distance travelled in the next 5 s can be found by splitting the shape into a triangle and a rectangle, and finding their areas separately.

The total distance travelled by the object in Figure E is the sum of the areas.

total distance travelled = 50 m + 50 m + 75 m = 175 m

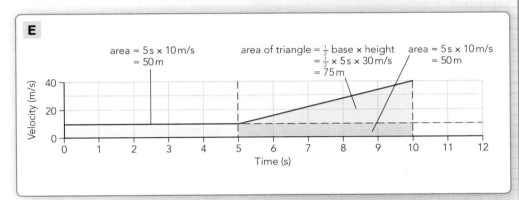

E

area = 5 s × 10 m/s
= 50 m

area of triangle = ½ base × height
= ½ × 5 s × 30 m/s
= 75 m

area = 5 s × 10 m/s
= 50 m

H **4 a** Calculate the distance travelled in each part of the graph (A–D) in Figure C.
H **b** What is the total distance travelled by the train?

5 Table F gives some data for a run of the *ThrustSSC*. Draw a velocity-time graph from this. Label your graph with all the things you can tell from it. ✏

F

Time after start of run (s)	Speed (m/s)
0	0
4	40
16	280
30	380
35	330
85	0

3 Use the graph in Figure C to calculate the acceleration of the train in each of parts A to D.

ResultsPlus
Watch Out!

Students often confuse distance-time graphs and velocity-time graphs. Make sure you read the axes carefully and take note of the units.

Skills spotlight

Scientists work with data in different forms. What are the advantages and disadvantages of using numbers rather than graphs to compare accelerations?

Learning Outcomes

3.6 Interpret velocity/time graphs to:
  a compare acceleration from gradients qualitatively
  b calculate the acceleration from the gradient (for uniform acceleration only)
  H c determine the distance travelled using the area between the graph line and the time axis (for uniform acceleration only)
HSW **10** Use qualitative and quantitative approaches when presenting scientific ideas and arguments, and recording observations

>>>>>>>>>>>>>>>>>>>>>>>>>>> Why do astronauts spend a lot of time underwater? **227**

What is an action-reaction pair?

When standing on the ground you feel your weight in your feet because the Earth is pushing back on your feet. Astronauts train underwater because the pushing force from the water (upthrust) is spread all over their bodies, making them feel lighter - similar to the feeling of being in space and having no pushing force.

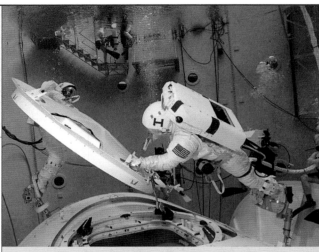
A Astronauts train underwater, because moving in space is a little like moving underwater.

Skills spotlight

Scientists use diagrams to make problems easier to analyse. How does a free-body diagram make it easier to understand the forces acting on an object?

A **force** is a push or a pull. Forces can make objects change speed, shape or direction. A force is a vector quantity because it has both size and direction. Most objects have more than one force acting on them. A **free-body diagram** shows the different forces acting on a single object, with their size and direction.

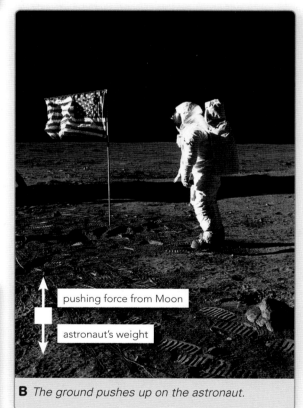

pushing force from Moon

astronaut's weight

B The ground pushes up on the astronaut.

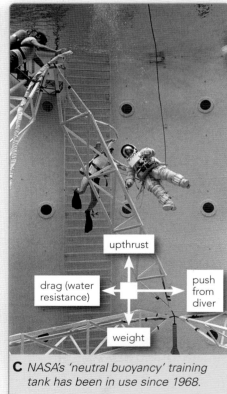

upthrust

drag (water resistance)

push from diver

weight

C NASA's 'neutral buoyancy' training tank has been in use since 1968.

1 Look at the free-body diagram in Figure C.
a Name two forces that act in opposite directions.
b Which force is larger, the push from the diver or the drag force? How do you know?

In Figure C, the forces acting on the astronaut are his weight, the upthrust from the water, the push to the right from the diver and the drag from the water that opposes his sideways movement.

Action and reaction forces

Whenever two objects touch they **interact** with each other. This interaction involves two forces, acting on *different* objects.

When a footballer heads a ball, they exert a force on the ball, called the **action force**. The ball exerts an equal and opposite force on the footballer, called the **reaction force**. This pair of forces is called an action-reaction pair.

Action and reaction forces are used to propel rockets. Gases are pushed out from the rear of the rocket. This creates the action force. The reaction force from the gases pushes the rocket forwards.

reaction force of ball

action force of head

D *The forces in an action-reaction pair always act on two different objects.*

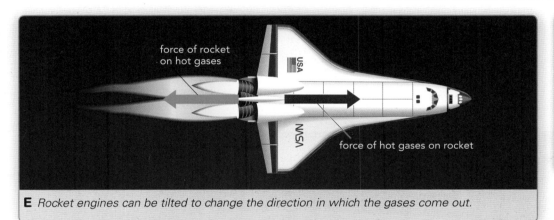

force of rocket on hot gases

force of hot gases on rocket

E *Rocket engines can be tilted to change the direction in which the gases come out.*

2 A bird sits on a fence post. The forces on the bird are its own weight and the reaction force from the fence post. Draw the free-body diagram for the bird.

3 What forces make up the action-reaction pair in Figure E?

4 If a rocket pushes gases directly behind it with a force of 53 000 N, what force pushes the rocket forwards?

5 What would happen to the force on the rocket in Figure E if the gases were pushed out in a different direction?

6 Draw free-body diagrams for the footballer and the ball in Figure D. Remember that a free-body diagram only shows the forces acting on a single object.

7 The astronaut in Figure C pushes against the diver. Draw and label a diagram to show the pair of action and reaction forces in this interaction. How is your diagram different from a free-body diagram?

ResultsPlus
Watch Out!

Students get confused when using the term 'reaction' force. Remember that action and reaction forces act on *different* objects when they meet.

Learning Outcomes

3.7 Draw and interpret a free-body force diagram

3.8 Demonstrate an understanding that when two bodies interact, the forces they exert on each other are equal in size and opposite in direction and that these are known as action and reaction forces

HSW *11* Present information using scientific conventions and symbols

How far could a space probe travel with no fuel?

What is a resultant force?

This is a hypersonic aeroplane that could travel at up to 4000 mph. The plane would carry 300 passengers and fly very high in the Earth's atmosphere (25 km) where the air is very thin.

A *A flight to Australia would take about five hours on this plane. However, there are no windows!*

> **1** In Figure B, how can you tell that drag and thrust are unbalanced?
>
> **2** Draw a free-body diagram for the aeroplane in Figure B.

Aeroplane designers think a lot about forces. Figure B shows some forces on the aeroplane. The force of gravity (the plane's weight) pulls it towards Earth. A force called 'lift' is produced by the wings and pushes the plane up. Forces that are equal but act in opposite directions are 'balanced'. In Figure B, lift and weight are balanced.

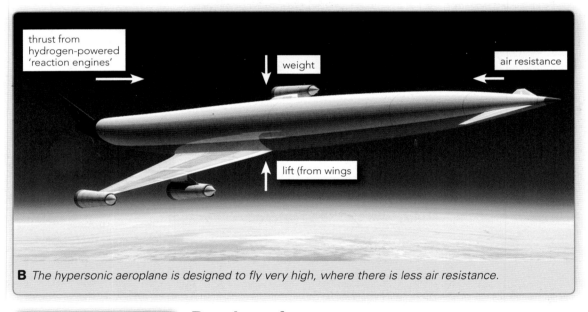

thrust from hydrogen-powered 'reaction engines'

weight

air resistance

lift (from wings

B *The hypersonic aeroplane is designed to fly very high, where there is less air resistance.*

'Thrust' is a force produced by the engines to push the aeroplane forwards. As it flies, air molecules push against it producing an opposite force called **air resistance** (a type of **drag**). In Figure B, the thrust is greater than the air resistance. The forces are unbalanced.

Resultant force

> **3** A boat has 1000 N of thrust. It is slowed by air resistance (300 N) and friction between the boat and the water (700 N). What is the resultant force?

When more than one force acts on an object, the effect is the same as if they were combined into a single force – the **resultant** force. If the forces act in the same direction, we add them together. If they act in opposite directions we subtract them.

> **e.g.**
>
> For Figure B:
> total thrust force to right = 100 000 N
> total drag force to left = 25 000 N
>
> resultant force = 100 000 − 25 000 = 75 000 N of force to the right.

Resultant force and acceleration

If the resultant force on an object is zero then the movement of the object does not *change*. If it is stationary, it will remain stationary. If it is moving, it will carry on moving at a constant velocity (in a straight line).

If the resultant force on an object is not zero, the object will accelerate in the direction of the resultant force.

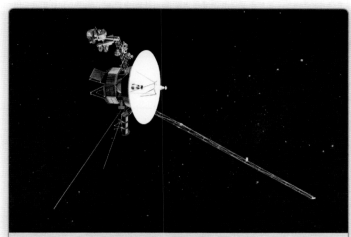

C *Voyager 1 doesn't need engines to keep it moving and has travelled over 10 billion miles so far. There is no air in space to slow it down.*

D *With no thrust from the engines, Virgin Galactic's SpaceShipTwo has a negative acceleration (slows down) as it comes in to land.*

7 Look at Figure D. Explain, using the idea of resultant forces, why SpaceShip Two is losing forward speed and height. ✎

4 Jane, Sameer and Kent are pushing a car. Jane pushes with 50 N, Sameer with 45 N and Kent with 60 N. Friction between the tyres and the road is 50 N and air resistance is 2 N. What is the resultant force?

5 Look at Figure B.
a Why will the aeroplane remain at a constant height?
b Why is the aeroplane accelerating?

6 Why does Voyager 1 not need engines on its journey through space?

Skills spotlight

Scientists look at things both quantitatively (using numbers) and qualitatively (not using numbers). Identify an example of each approach on this page.

Learning Outcomes

3.9 Calculate a resultant force using a range of forces (limited to the resultant of forces acting along a line) including resistive forces

3.10 Demonstrate an understanding that if the resultant force acting on a body is zero, it will remain at rest or continue to move at the same velocity

3.11 Demonstrate an understanding that if the resultant force acting on a body is not zero, it will accelerate in the direction of the resultant force

HSW *10* Use qualitative and quantitative approaches when presenting scientific ideas and arguments, and recording observations

P2.13 Investigating force, mass and acceleration

 What is the relationship between force, mass and acceleration?

If you go ten pin bowling, you may be able to throw the ball with a speed of around 8 m/s. A top cricketer can throw a cricket ball with a speed of up to 45 m/s. Why are the two top speeds so different?

The way you throw a ball depends on its size and mass. The top speed of a ball depends on the acceleration you give it while it is in contact with your hand.

Once a cricket ball leaves your hand the only forces on it are air resistance, gravity and friction. Gravity pulls the cricket ball downwards as it flies through the air, so it accelerates the ball downwards. Gravity does not affect its forward speed.

Gravity does not affect the path of a bowling ball because it runs along the lane. If the bowling alley were long enough, the ball would eventually come to a stop because of friction and air resistance. The effects of friction could be compensated for by having the bowling alley on a slight slope. If the angle of the slope is just right, gravity pulling the ball down the slope will just cancel out the forces of friction trying to make the ball slow down.

A *A bowling ball moves at speeds of up to 8 m/s.*

You can use trolleys and ramps to investigate force, mass and acceleration. The angle of the ramp can be adjusted to cancel out the effects of friction.

Your task

You are going to plan an investigation that will allow you to find out how the mass of an object affects its acceleration. Your teacher will provide you with some materials to help you organise this task.

Learning Outcomes

3.15 Investigate the relationship between force, mass and acceleration

When planning an investigation like this, one of the skills you will be assessed on is your ability to *choose suitable equipment*. There are 2 marks available for this skill. Here are two extracts focusing on this skill. Other skills are dealt with in other lessons.

| Student extract 1 | A basic response for this skill |

The student describes one piece of apparatus but explains why they need something else.

> I will use a trolley running down a ramp. I need the weights to pull on the trolley.

This is describing some apparatus that the student will use. However, it would be better to also explain why the trolly will be needed.

| Student extract 2 | A good response for this skill |

Explain why each piece of equipment is needed as you list it.

> I will use a trolley on a ramp, as the wheels will help to reduce the effects of friction. I will pull the trolley using masses on a string going over the pulley, as this will allow me to apply a constant force to the trolley. I will use light gates to measure the speed at two points and the time taken to go between the two points, as I can use this information to calculate the acceleration. Light gates and a computer will provide better quality data than I could obtain using a stopclock.

This is a good response as the student has described each piece of equipment needed *and* explained why it is needed.

To access 2 marks

- Choose the most relevant equipment and resources
- Explain the reasons for your choices and why they are relevant to your method

How is acceleration related to the size of a force?

Rockets and racing bikes both need to be strong and as light as possible. Racing cyclists need to get as much acceleration as they can from the force they apply to the pedals. Research into strong, light materials for use in space can give us new materials that we can put to other uses, including making faster bikes for the future.

A *Nicole Cooke winning Gold for the road race at the Beijing Olympics in 2008. Her bike had a mass of just 6.8 kg.*

1 The resultant force on an object is not zero. What will happen to the object?

2 The same force is used to throw a tennis ball and a cricket ball. What will be different about their motions? Explain your answer.

3 If you wanted to make a small car and a large lorry accelerate at the same rate, what can you say about the forces needed to do this? Explain your answer.

When there is a resultant force acting on an object, the object will accelerate in the direction of the resultant force. The size of that acceleration will depend upon two things:
- the size of the force (for the same mass, the bigger the force the bigger the acceleration)
- the mass of the object (for the same force, the more massive the object the smaller the acceleration).

If the same force is applied to two objects, the one with the smaller mass will accelerate more. If two objects of different mass are to have the same acceleration, then a larger force must be applied to the object with the greater mass.

B *Two very different catapults*

4 The catapults in Figure B both fire a rock of the same mass. Which will fire it faster and why?

Calculating forces

The force needed to accelerate a particular object can be calculated using the equation:

force (N) = mass (kg) × acceleration (m/s²)

This is often written as $F = m \times a$
A newton is the force needed to accelerate a mass of 1 kg by 1 m/s².

The shorter the time it takes a car to reach 60 mph from rest, the larger the acceleration. The small car in Figure C accelerates three times as quickly as the larger car.

C Both cars have the same engine.

Maths skills

Rearranging simple equations like $F = ma$ can be done using a formula triangle, as shown in Figure E in P2.3.

$$F = m \times a$$

$$a = \frac{F}{m}$$

$$m = \frac{F}{a}$$

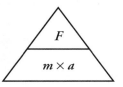
F

$m \times a$

e.g. The larger car in Figure C has an acceleration of 3.0 m/s². What is the force provided by the engine?

$$F = m \times a$$
$$F = 1500\,kg \times 3\,m/s^2$$
$$F = 4500\,N$$

ResultsPlus
Watch Out!

Students can lose marks by using the wrong units. To find the force in newtons, the mass must be in kg and the acceleration in m/s².

? 5 The smaller car in Figure C has an acceleration of 9.0 m/s². What is the force provided by its engine?

H 6 A cheetah of mass 40 kg can produce a force of 360 N to push itself forwards. What acceleration does this give the animal?

7 Explain in terms of mass, force and acceleration why racing cyclists and their bikes need to be as light as possible to achieve the greatest possible acceleration at the start of a race.

Skills spotlight

Scientists use symbols to represent things. Write down the symbols that are used in the equation that relates force, mass and acceleration and what they stand for.

Learning Outcomes

3.12 Demonstrate an understanding that a resultant force acting on an object produces an acceleration which depends on:
a the size of the resultant force
b the mass of the object

3.13 Use the equation:
force (newton, N) = mass (kilogram, kg) × acceleration (metre per second squared, m/s²)

$$F = m \times a$$

HSW **11** Present information using scientific conventions and symbols

>>>>>>>>>>>>>>>>>>>>>> When would a hammer and a feather both fall in the same way?

What is terminal velocity?

In 1960, Joseph Kittinger jumped from the gondola of a balloon at a height of 31 333 m. It took him 13 minutes and 45 seconds to reach the ground, accelerating to a maximum speed of 988 km/h (614 mph) when he was in 'free fall'.

A *Joseph Kittinger jumping in 1960. A sign by the door on the gondola read 'This is the highest step in the world'.*

1 What is the difference between mass and weight?

2 A landing craft of mass 300 kg goes to Jupiter, where the gravitational field strength is 25 N/kg.
a What would its mass be?
b What would its weight be?

Mass is the quantity of matter there is in an object and does not change, but **weight** is a measure of the pull of gravity on an object and can change. To describe how strong gravity is we talk about **gravitational field strength**, measured in newtons per kilogram (N/kg). On Earth this has a value of about 10 N/kg. This means that each kilogram is pulled down with a force of 10 N. The weight of any object on Earth can be calculated using the following equation.

weight (N) = mass (kg) × gravitational field strength (N/kg)

This is often written as: $W = m \times g$

e.g.

What is the weight of a 300 kg planetary landing craft on the surface of the Earth?

$$W = m \times g$$
$$W = 300\,kg \times 10\,N/kg = 3000\,N$$

ResultsPlus
Watch Out!

Students can confuse force and acceleration when answering questions about falling objects. In a vacuum, all falling objects accelerate at the same rate but that does not mean that the forces are the same.

Larger masses need more force to get them to accelerate than smaller masses. As mass increases, so does the force from gravity (weight). This means that gravity accelerates all objects down towards the Earth by the *same* amount. So a feather and a hammer dropped together should accelerate by the same amount and so hit the ground at the same time. They do not do this on Earth because there is more air resistance on the feather, which slows it down.

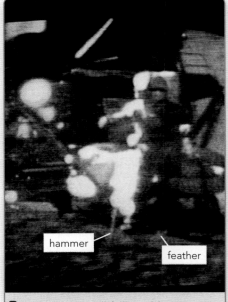

B *In a vacuum, all falling objects accelerate at the same rate.*

3 Why did the hammer and feather hit the ground together when dropped on the Moon?

On the Moon in 1971, astronaut David Scott tried this experiment. The feather and the hammer both hit the ground at the same time because they were both dropped in a vacuum.

Terminal velocity

When an object just starts to fall from above the Earth, the only force acting on it is its own weight, which accelerates it towards the Earth. As it gains speed, air resistance begins to act in the opposite direction, reducing the resultant force.

Figure C shows how the resultant force on a falling skydiver changes as their velocity increases. As with all objects, eventually the weight and air resistance forces are balanced. The resultant force on the skydiver is then zero, so they will stop accelerating. They will travel at a constant velocity, called the **terminal velocity**.

Just after jumping, there is little air resistance.

As speed increases, so does air resistance.

Eventually weight and air resistance are the same.

Large resultant force downwards, the skydiver accelerates.

Smaller resultant force downwards, the skydiver still accelerates.

Resultant force is zero, terminal velocity reached.

C *Eventually a skydiver reaches a terminal velocity.*

4 What is the name for the maximum velocity any object falling in air can reach?

5 What happens to the size of the air resistance force on an object as its speed increases?

6 a What is the weight of a skydiver of mass 60 kg?
b In what direction does this force act?
c What is the force of air resistance when this skydiver is at their terminal velocity?
d In what direction does this force act?

7 Two identical crates are dropped from an aircraft. One is empty and the other is full. Which crate will reach the ground first? Explain your answer.

Skills spotlight

Scientists plan how to test ideas. How could you use a stopclock and a ruler to find whether paper cupcake cases reach a terminal velocity when dropped?

Learning Outcomes

3.14 Use the equation:
weight (newton, N) = mass (kilogram, kg) × gravitational field strength (newton per kilogram, N/kg)
$W = m \times g$

3.16 Recall that in a vacuum all falling bodies accelerate at the same rate

3.17 Demonstrate an understanding that:
a when an object falls through an atmosphere air resistance increases with increasing speed
b air resistance increases until it is equal in size to the weight of the falling object
c when the two forces are balanced acceleration is zero and terminal velocity is reached

HSW 5 Plan to test a scientific idea by controlling relevant variables

Stopping distances

 What affects the distance covered when a vehicle tries to stop?

In 1971, Dave Scott of the Apollo 15 Moon mission said that driving the Moon buggy was 'quite good… the steering is responsive… and we average eight kilometres per hour'. NASA called them Lunar Rover Vehicles but because it was like driving a sand kart known as a dune buggy, these vehicles were nicknamed Moon buggies.

A *Driving a Moon buggy was like driving a go-kart with very bouncy suspension.*

?

1 Why is it important for a driver to know their stopping distance?

2 A go-kart driver has a thinking distance of 5 m and a braking distance of 12 m. What is the overall stopping distance?

Stopping safely

Go-kart drivers need to know what affects the distance it takes for a kart to stop safely in an emergency.

When a driver sees a problem ahead, the go-kart will travel some distance whilst the driver reacts to seeing the problem. This is called the **thinking distance**. The kart will then go some distance further whilst the brakes are working to bring it to a halt. This is called the **braking distance**. The overall **stopping distance** for any road vehicle will be the sum of the thinking and braking distances.

B *If there is a crash on a go-karting circuit, the next kart along has nowhere to go – it must stop.*

stopping distance = thinking distance + braking distance

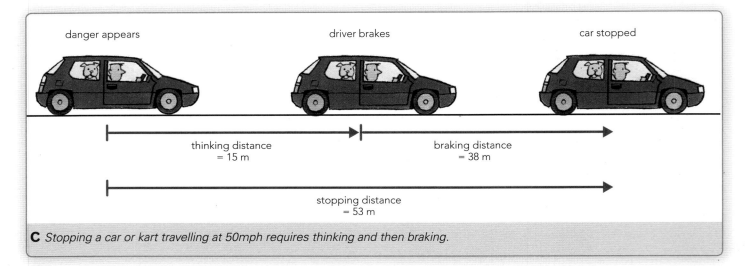

danger appears

driver brakes

car stopped

thinking distance
= 15 m

braking distance
= 38 m

stopping distance
= 53 m

C *Stopping a car or kart travelling at 50mph requires thinking and then braking.*

Changing stopping distances

A driver's thinking distance will be longer if their **reaction time** is greater – for example, if they are tired, or ill, or have taken drugs or alcohol.

If a vehicle is moving at a greater speed, then both the thinking distance and the braking distance are increased.

The force of **friction** slows vehicles down. Brakes create a lot of friction. If the brakes are worn, they create less friction and do not slow the vehicle so well. Friction between the tyres and road is also important. If the road is wet or has loose gravel on it, there is less friction and the braking distance is increased.

If a vehicle has more mass, more force is needed to decelerate it. So, if the same amount of friction is used to stop a vehicle, a heavier vehicle will travel further than a lighter one (it has a greater braking distance).

Skills spotlight

Science provides evidence that is used in making decisions about social issues. Write a scientific explanation as to why there are now many 20 mph zones in residential areas.

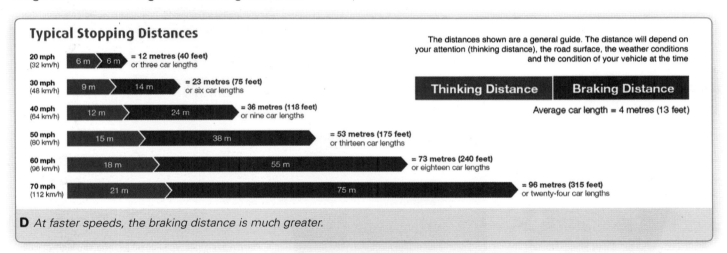

Typical Stopping Distances

20 mph (32 km/h)	6 m > 6 m	**= 12 metres (40 feet)** or three car lengths
30 mph (48 km/h)	9 m > 14 m	**= 23 metres (75 feet)** or six car lengths
40 mph (64 km/h)	12 m > 24 m	**= 36 metres (118 feet)** or nine car lengths
50 mph (80 km/h)	15 m > 38 m	**= 53 metres (175 feet)** or thirteen car lengths
60 mph (96 km/h)	18 m > 55 m	**= 73 metres (240 feet)** or eighteen car lengths
70 mph (112 km/h)	21 m > 75 m	**= 96 metres (315 feet)** or twenty-four car lengths

The distances shown are a general guide. The distance will depend on your attention (thinking distance), the road surface, the weather conditions and the condition of your vehicle at the time

Thinking Distance	**Braking Distance**

Average car length = 4 metres (13 feet)

D *At faster speeds, the braking distance is much greater.*

3 Lorry drivers have the same thinking distance as car drivers but their braking distance is greater. Why are the overall stopping distances for cars less than for lorries?

4 How would ice on the road affect thinking distance and braking distance?

5 Why is drinking-and-driving illegal?

6 The *Highway Code* is a government book about driving. Prepare a summary for a new version of the *Highway Code* explaining what things increase and decrease the stopping distance and why.

Results Plus
Watch Out!

Students often confuse reactions and reaction time. Faster reactions means shorter reaction time.

Learning Outcomes

4.1 Recall that the stopping distance of a vehicle is made up of the sum of the thinking distance and the braking distance

4.2 Demonstrate an understanding of the factors affecting the stopping distance of a vehicle, including
a the mass of the vehicle
b the speed of the vehicle
c the driver's reaction time
d the state of the vehicle's brakes
e the state of the road
f the amount of friction between the tyre and the road surface

HSW 13 Explain how and why decisions about uses of science and technology are made

P2.17 Friction with different surfaces

How is friction affected by the surfaces involved?

Ice biking is much more difficult than ordinary cycling as the riders have much less control. The lack of friction on ice makes steering and braking very difficult.

Ice is well known for being slippery. This can lead to fun activities like ice bike racing. On an icy surface, different tyres will have different amounts of grip. This will affect how easy the bike is to control.

A *Low friction makes for difficult riding conditions.*

B *If the ground surface is very smooth, like ice, then bike tyres need more grip.*

Your task

You are going to plan an investigation that will allow you to find out what effect different surfaces have on the amount of friction. Your teacher will provide you with some materials to help you organise this task.

Learning Outcomes

4.3 Investigate the forces required to slide blocks along different surfaces, with differing amounts of friction

ResultsPlus
Build Better Answers

When planning an investigation like this, one of the skills you will be assessed on is your ability to *write a hypothesis*. There are 4 marks available for this skill. Here are two extracts focusing on this skill. Other skills are dealt with in other lessons.

Student extract 1 | A basic response for this skill

This response includes a limited hypothesis, e.g. a prediction of what will happen in the investigation.

> The biggest friction force will be found with sandpaper under the block, then the cardboard surface, and the lowest friction will be with the wooden desk surface. This is because rougher surfaces have more friction.

The student has given a partial reason to back up the hypothesis.

Student extract 2 | A good response for this skill

This hypothesis covers all the parts of the experiment, and the hypothesis is based on scientific ideas.

> The friction force results will be in the same order as the order of roughness of the surfaces I test because rougher surfaces catch on each other more, creating more friction. If you look at any surface under a microscope, it is very uneven. The more jagged the surface, the more the movement of the two surfaces past each other is blocked, increasing the friction force.

The hypothesis is justified by more scientific details.

 ResultsPlus

To access 2 marks
- Provide a hypothesis that is appropriate for most of the task
- Try to justify your hypothesis

To access 4 marks
- Provide a hypothesis that is appropriate for the full scope of the task
- Justify your hypothesis using relevant scientific ideas

What is momentum?

To create a monster truck, a normal vehicle is fitted with a powerful engine, strong suspension and huge oversized wheels. Apart from being a very bouncy ride, drivers find them difficult to control, as all the truck's movements are oversized too.

A A small car can become a monster.

In a race, if a vehicle goes out of control and hits a barrier, the barrier moves. If two vehicles are going at the same speed and hit a barrier, the barrier moves more when hit by a monster truck than by a small car. The reason for this is because the larger vehicle has more **momentum**.

1 Calculate the momentum of:
a a 730 kg Smart car moving north at 20 m/s
b a 2500 kg monster truck moving at a velocity of 8 m/s south.

Skills spotlight

Scientists use knowledge of physics laws to analyse real situations and develop solutions to problems. Look at Figure B and explain how car designers could monitor the changes in momentum of the crashing car. How could they make quantitative measurements, how could they make qualitative measurements and what is an advantage of each?

B Much more damage would have been caused to the crash barrier if it had been hit by a heavier vehicle, such as a truck.

Momentum is a measure of how strongly something is moving. It is calculated by multiplying mass by velocity.

$$\text{momentum} = \text{mass} \times \text{velocity}$$
(kilogram metre per second, kg m/s) = (kilogram, kg) × (metre per second, m/s)

2 Why would a stationary monster truck have no momentum?

H 3 A meteorite flies through space at 250 m/s towards Earth. It has a momentum of 30 000 kg m/s towards Earth. What is its mass?

> **e.g.**
> The momentum of the monster truck in Figure A (mass = 4500 kg) when it is travelling at 12 m/s east would be:
>
> momentum = mass × velocity = 4500 × 12 = 54 000 kg m/s east

Notice that since velocity is a **vector quantity** (it has a size *and* direction), so is momentum. So we should always state the amount of momentum an object has and the direction of that momentum, which will be the same direction as its velocity.

Conservation of linear momentum

When a moving object collides with another object, the *total* momentum of both objects is the same before the collision as it is after the collision. This is known as conservation of linear momentum. A crash barrier isn't moving and so has zero momentum. If a vehicle hits it, the barrier will start to move. The velocity of the barrier and the vehicle together is less than the vehicle on its own (because the combined mass is greater). However, the total momentum remains the same.

If you find the total momentum before a collision and then the total momentum after the collision, they will be the same. Remember, momentum is a vector so you need to consider direction when you 'add' the quantities together.

C *Some of the car's momentum is transferred to the ball.*

4 a Calculate the momentum of each penguin before they collide.
b Calculate the total momentum before the penguins collide.
c Say which direction the total momentum is in before the collision.
d Calculate the total momentum after the collision and say in which direction it is.
e Has the momentum been conserved?

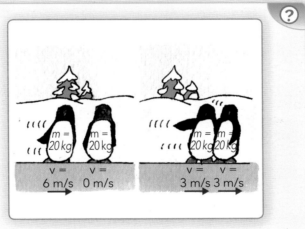

m = 20 kg m = 20 kg
v = 6 m/s v = 0 m/s

m = 20 kg m = 20 kg
v = 3 m/s v = 3 m/s

5 Put the following things in order of increasing momentum and explain how you worked out the order: a monster truck in a race; a running dog; a sleeping cat; the meteorite from question **3**.

6 Think about a car crashing. Write an explanation of how to work out its momentum and explain what is meant by conservation of linear momentum.

Learning Outcomes

4.4 Use the equation:
momentum (kilogram metre per second, kg m/s) = mass (kilogram, kg) × velocity (metre per second, m/s) to calculate the momentum of a moving object

4.5 Demonstrate an understanding of momentum as a vector quantity

4.6 Demonstrate an understanding of the idea of linear momentum conservation

HSW **10** Use qualitative and quantitative approaches when presenting scientific ideas and arguments, and recording observations

>>>>>>>>>>>>>>>>>>>>>> How do crash test dummies tell us when they are hurt? 243

P2.19 Crumple zones

How can we reduce the forces on a crashing car?

New cars are tested to see how safe they are in a crash, and are given a score. Crash test dummies are used to model the people. The dummies sometimes have different coloured paints applied to different areas to show which parts of a person's body will hit which part of the inside of the car.

A *Crash test dummies have force sensors which are linked to data logging computers.*

Cars contain many safety features, like air bags and **crumple zones**. If you look at the car in Figure A you will see that the front of the car has crumpled. The car was designed to do this. The crumpling of the metal means that the deceleration of the people in the car is less, and so there is less force on them. This means they are less likely to be injured.

B *This trolley has been set up to model crashes.*

In school, we could model crash testing. This can be filmed for analysis afterwards, or you can use sensors attached to a computer for data logging measurements like speed and force. Like the real scientists though, we could also think low-tech. Soft materials like plasticine will deform when hit, and so looking at the dents in the plasticine can show how hard it was hit.

Your task

You are going to plan an investigation that will allow you to find out how crumple zones can be used to reduce the forces in collisions. Your teacher will provide you with some materials to help you organise this task.

Learning Outcomes

4.8 Investigate how crumple zones can be used to reduce the forces in collisions

Build Better Answers

When planning an investigation like this, one of the skills you will be assessed on is your ability to *control variables*. There are 6 marks available for this skill. Here are three extracts focusing on this skill. Other skills are dealt with in other lessons.

Student extract 1 — A basic response for this skill

The phrase 'fair test' is too general, you must 'control variables'.

> I will make it a fair test. I am going to make sure the impact is at the same speed every time, using a light gate, just before it hits the block.

You must identify a control *variable* – something that can be measured.

You need to say how to control this variable.

Student extract 2 — A better response for this skill

Identify as many variables as you can.

> The total mass of the vehicle and attached crumple zone , and the collision speed must both be kept the same each time. If I start with the heaviest design of crumple zone, then for each other design, I can add a Blu-Tack weight to make it up to the same as the first one. I will check the weight is always the same using a digital balance. The slide distance for each test run will be kept constant by releasing from a marked point on the ramp each time.

You must explain what will be measured to control the variables, and how it will be measured. You should not just say 'use the same equipment' or similar statements.

Student extract 3 — A good response for this skill

Each variable must be identified, and then you must explain how to control it.

> The total mass of the vehicle and attached crumple zone are weighed using a balance and must be kept the same each time. If I start with the heaviest design of crumple zone, then for each other design, I can add a Blu-Tack weight to make it up to the same as the first one. The slide distance for each test run will be kept constant by releasing from a marked point on the ramp each time, so the speed is always the same at impact. Mass and velocity will affect the momentum in the collision so must be kept constant. The front end surface area of crumple zone will be kept constant, by using a ruler when building it, so that the design shape is not a factor in the crumpling force involved. The ramp angle will be maintained by keeping the top and bottom ends at the same heights each time. This will also keep the speed the same every time.

You must explain *why* it is important to control the variables you suggest, as well as *how* to control them.

How is momentum used in the design of vehicle safety devices?

Figure A shows an internal air bag jacket for motorcyclists. It contains compressed gas. If the rider comes off the bike in an accident, a trigger releases the gas to fill the bag, protecting the rider.

A Air bags reduce the forces acting on people in collisions.

When you are travelling in a car, your body is going at the same speed as the car. If the car stops suddenly, your momentum continues to carry you forward. If your body is stopped very suddenly, by hitting the dashboard for example, the huge forces can cause very serious injury.

The faster the momentum of an object changes, the greater the force on the object. So vehicle safety measures try to reduce the rate of change of momentum to reduce the forces.

B Your momentum carries you forward in a car accident.

In an accident, a **seat belt** stretches. This reduces the rate of change of momentum and so reduces the force on the passenger. However, in high-speed accidents the force is so great that the seat belt itself may cause injury.

1 Why do vehicle safety measures try to reduce momentum over a long period of time?

2 Why is it more dangerous for your head to hit the car dashboard than for it to hit an air bag?

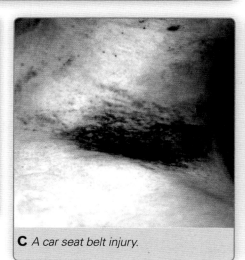

C A car seat belt injury.

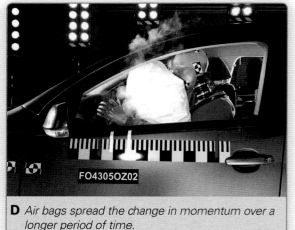

D Air bags spread the change in momentum over a longer period of time.

Cars are now fitted with **air bags**, which reduce momentum to zero more slowly and help prevent seat belt injuries. However, they only inflate at speeds of more than 10–15 mph and so do not replace seat belts.

Cars also have **crumple zones**. In an accident, the material in these crumple zones squashes and folds in a specific way to reduce the momentum of the car over a longer period of time.

E *Crumple zones increase the time taken to stop a car.*

H

Changing momentum

A force is needed to change the momentum of an object. The rate of change of the momentum is equal to the force applied. The force can be worked out using this equation:

$$\text{force (newton, N)} = \text{rate of change of momentum}$$

$$= \frac{\text{change in momentum (kilogram metre per second, kg m/s)}}{\text{time taken for the change (second, s)}}$$

$$F = \frac{mv - mu}{t}$$

Here, u is the initial velocity and v is the final velocity, so $mv - mu$ is the change in momentum.

e.g.

What is the force needed to change a cyclist's momentum by 24 kg m/s in 4 seconds?

$$F = \frac{\text{change in momentum}}{\text{time}} = \frac{24}{4} = 6\,\text{N}$$

?

H 5 Calculate the following:
a What is the force needed to change a car's momentum by 3000 kg m/s in 5 seconds?
b The momentum of a boat changes from 120 kg m/s to 320 kg m/s in 5 seconds. What is the rate of change of momentum? What is the force needed?

H 6 A 75 kg man starts a sprint race. He increases his velocity from rest to 6 m/s using a force of 200 N. How long did it take him to reach 6 m/s?

7 List three vehicle safety devices. For each one write a brief explanation of how it can reduce injuries in a traffic accident. 🖉

3 'Bubble wrap' is plastic covering with many air bubbles. How do you think bubble wrap protects fragile items?

4 Think about an empty egg carton. How is a car crumple zone:
a similar to an egg carton
b different from an egg carton?

Skills spotlight

Scientists need to be able to plan tests for new devices to show how they will perform. Write a plan for an experiment to find out how many layers of bubble wrap are needed to protect an egg when dropped from a fixed height.

Maths skills

H To change the subject of a formula, carry out the same operations on both sides of the equals sign until the subject is on it's own on one side of the equals sign.
E.g. To make v the subject of $F = \dfrac{mv - mu}{t}$ you can:
1. multiply both sides by t so $Ft = mv - mu$
2. add mu to both sides so $Ft + mu = mv$
3. divide both sides by m so
$$v = \frac{Ft + mu}{m} \text{ or}$$
$$v = \frac{Ft}{m} + u$$

Learning Outcomes

4.7 Demonstrate an understanding of the idea of rate of change of momentum to explain protective features including bubble wraps, seat belts, crumple zones and air bags

H 4.9 Use the equation:
force (newton, N) = change in momentum (kilogram metre per second, kg m/s) / time (second, s)
$F = (mv - mu) / t$
to calculate the change in momentum of a system, as in 4.6

HSW 5 Plan to test a scientific idea, answer a scientific question, or solve a scientific problem by controlling relevant variables

What do we mean by power?

Motorbike licences have different categories depending on the power the engine can produce.

A Motorbikes have different power outputs.

The physics meaning of 'work'

Work is done when energy is transferred from one form to another.

Work is an amount of **energy transferred** so it is measured in **joules (J)**. The amount of work done is equal to the amount of energy transferred.

B Car brakes 'do work' by transferring kinetic energy to thermal energy (heat). Work can be done in many different ways.

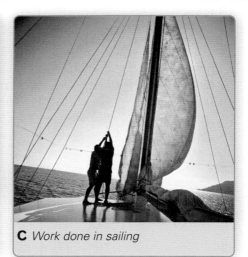

C Work done in sailing

Calculating work

work done = force × distance moved (in the direction of the force)

$$E \quad = \quad F \quad \times \quad d$$

(joules, J) = (newtons, N) × (metres, m)

e.g.

A sailor lifts a 300 N sail from the deck to a height 4 m up. How much work does the sailor do?

work done = force × distance moved (in the direction of the force)

$$E \quad = \quad F \quad \times \quad d$$
$$E \quad = \quad 300 \quad \times \quad 4$$
$$E \quad = \quad 1200 \text{ joules}$$

Power

Power is the rate of doing work. It is measured in **watts (W)**. 1 watt means 1 joule of work done per second.

$$\text{power (watt, W)} = \frac{\text{work done (joule, J)}}{\text{time taken (second, s)}}$$

$$P = \frac{E}{t}$$

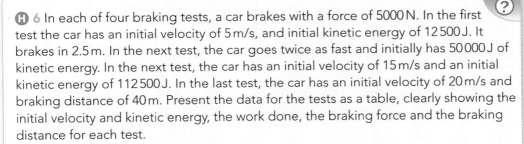

e.g.

A motorbike engine increases the bike's **kinetic energy** over 40 m, using a force of 6000 N. It takes 5 seconds to do this. What power did the engine provide?

$$E = F \times d = 6000 \times 40 = 240\,000\,\text{J}$$

$$P = \frac{\text{work done}}{\text{time taken}} = \frac{E}{t} = \frac{240\,000}{5} = 48\,000\,\text{W}$$

?

4 From question 3, calculate the power in each case if the times taken were:
a 2 seconds
b 400 s
c 8 s.

5 Explain why the breaking distance is greater for a greater speed.

Braking distances and work [H]

The work done stopping a vehicle is the braking force multiplied by the braking distance. When stopped, all of the vehicle's kinetic energy has been removed. So, the work done is equal to the initial kinetic energy.

?

H 6 In each of four braking tests, a car brakes with a force of 5000 N. In the first test the car has an initial velocity of 5 m/s, and initial kinetic energy of 12500 J. It brakes in 2.5 m. In the next test, the car goes twice as fast and initially has 50000 J of kinetic energy. In the next test, the car has an initial velocity of 15 m/s and an initial kinetic energy of 112500 J. In the last test, the car has an initial velocity of 20 m/s and braking distance of 40 m. Present the data for the tests as a table, clearly showing the initial velocity and kinetic energy, the work done, the braking force and the braking distance for each test.

7 Brakes that overheat can fail. Write a short leaflet for a company that makes brake parts, to tell people why brakes get hot and why braking more suddenly might make them get even hotter.

Skills spotlight

Power outputs are sometimes compared using figures (quantitatively) and sometimes just ranked in order (qualitatively). Suggest one situation in which each type of comparison would be more appropriate than the other.

Learning Outcomes

4.10 Use the equation:
work done (joule, J) = force (newton, N) × distance moved in the direction of the force (metre, m) $E = F \times d$

4.11 Demonstrate an understanding that energy transferred (joule, J) = work done (joule, J)

4.12 Recall that power is the rate of doing work and is measured in watts, W

4.13 Use the equation:
power (watt, W) = work done (joule, J) / time taken (second, s) $P = E / t$

4.14 Recall that one watt is equal to one joule per second, J/s

H 4.18 Carry out calculations on work done to show the dependence of braking distance for a vehicle on initial velocity squared (work done to bring a vehicle to rest equals its initial kinetic energy)

HSW 10 Use qualitative and quantitative approaches when presenting scientific ideas and arguments, and recording observations

How are kinetic and gravitational potential energies calculated?

When NASA deliberately crash landed the Rover Exploration Missions on Mars, they were travelling very fast. The delicate instruments were protected using air bags.

B Testing the Mars Rover air bag system involved giving it gravitational potential energy by lifting it with a crane.

Gravitational potential energy

Gravitational potential energy is energy that is stored because of an object's position in a gravitational field. Essentially, on Earth if something can fall it has gravitational potential energy. It can be calculated using this equation:

A The Mars Exploration Rover's gravitational potential energy was transferred into kinetic energy so it landed at high speed.

$$\text{gravitational potential energy (joule, J)} = \text{mass (kilogram, kg)} \times \text{gravitational field strength (newton per kilogram, N/kg)} \times \text{vertical height (metre, m)}$$

$$\text{GPE} = m \times g \times h$$

On Earth, the strength of the gravitational field is 10 N/kg.

> **e.g.**
>
> What is the gravitational potential energy gained by a 500 kg Mars Rover when it is lifted 15 m by a test crane on Earth?
>
> $$\text{GPE} = m \times g \times h$$
> $$= 500 \times 10 \times 15 = 75\,000\,\text{J}$$

1 Calculate the gravitational potential energy of the following:
a A 2 kg toy robot dog jumps up 1 metre.
b A lift has a mass of 400 kg and rises 24 m.
c The Mars Exploration Rover, 15 m above the surface of Mars, where the gravitational field strength is 4 N/kg.

Kinetic energy

Kinetic energy is the scientific name for movement energy. We can calculate the kinetic energy an object has using the equation:

$$\text{Kinetic energy (joule, J)} = \tfrac{1}{2} \times \text{mass (kilogram, kg)} \times \text{(velocity)}^2 \text{ (metre/second)}^2,\ \text{(m/s)}^2$$

$$\text{KE} = \tfrac{1}{2} \times m \times v^2$$

Watch Out!

Students often forget to square the velocity in kinetic energy calculations.

> **e.g.**
>
> What is the kinetic energy of a 65 kg girl running at 6 m/s?
>
> $$\text{KE} = \tfrac{1}{2} \times m \times v^2$$
> $$= \tfrac{1}{2} \times 65 \times 6^2 = \tfrac{1}{2} \times 65 \times 36$$
> $$= 1170\,\text{J}$$

Conservation of energy

When energy is transferred, the total amount always remains the same. So when an object with gravitational potential energy falls down, the amount of kinetic energy it has at the instant of hitting the ground is equal to its initial gravitational potential energy.

For example, when the Mars Exploration Rover was dropped in the test shown in Figure B, the 75 000 joules of gravitational potential energy were all transferred to kinetic energy by the time it hit the ground.

e.g.

How fast did the Mars Exploration Rover land in its air bag test?

$$KE = 75\,000\,J = \tfrac{1}{2} \times m \times v^2$$
$$= \tfrac{1}{2} \times 500 \times v^2$$
$$v^2 = \frac{75\,000}{(\tfrac{1}{2} \times 500)} = \frac{75\,000}{250} = 300$$
$$v = \sqrt{300} = 17.3\,m/s$$

Braking distances

We have seen that the braking distance for a car is much greater if its speed is faster. This is because the maximum braking force is constant for any given vehicle, but the amount of kinetic energy to be removed goes up with the square of the speed, since $KE = \tfrac{1}{2} mv^2$.

C This car's kinetic energy was changed suddenly into other forms of energy.

H

5 Why do brakes get hot when a car slows down?

6 How can a landing spacecraft's gravitational potential energy be calculated? How would its kinetic energy be calculated? Explain the energy transfers that happen.

2 Calculate the kinetic energy of the following:
a A 2 kg toy robot dog walks at 2 m/s.
b A boy on a bike rides at 8 m/s. The mass of the boy and his bike is 70 kg.
c A whale swims at 7 m/s. The whale's mass is 4000 kg.

H 3 The whale squirts 25 kg of water into the air with 1250 J of kinetic energy. How fast is the water moving initially?

H 4 A 55 kg girl jumps upwards at 2 m/s. If all her kinetic energy is transferred to gravitational potential energy, how high did she jump?

Skills spotlight

To store gravitational potential energy, we have to do work. Note down the equations for work done and for gravitational potential energy and write an explanation of why equations are useful to scientists.

Learning Outcomes

4.15 Use the equation:
gravitational potential energy (joule, J) = mass (kilogram, kg) × gravitational field strength (newton per kilogram, N/kg) × vertical height (metre, m)
$GPE = m \times g \times h$

4.16 Use the equation:
kinetic energy (joule, J) = ½ × mass (kilogram, kg) × velocity² ((metre/second)², (m/s)²)
$KE = \tfrac{1}{2} \times m \times v^2$

4.17 Demonstrate an understanding of the idea of conservation of energy in various energy transfers

H 4.18 Carry out calculations on work done to show the dependence of braking distance for a vehicle on initial velocity squared (work done to bring a vehicle to rest equals its initial kinetic energy)

HSW **11** Present information using scientific conventions and symbols

How can atoms of the same element be different?

P2.23 Isotopes

What is an isotope?

During missions to the Moon in the 1960s and 1970s, 381 kg of Moon rocks and soil were brought back to Earth. Scientists can find the ages of these rocks by comparing different types of atoms (called isotopes) in the rocks.

A This piece of moon rock is 3.9 billion years old.

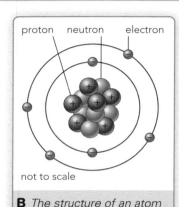

not to scale

B The structure of an atom

Atoms contain electrons, protons and neutrons, which are called **sub-atomic particles**. Protons and neutrons are in the nucleus and electrons surround the nucleus. The particles found in the nucleus are called **nucleons**.

All atoms of a particular element have the same number of protons. This number is called the **atomic number** or **proton number** of the element. Atoms of different elements have different numbers of protons and so have different atomic numbers.

The number of neutrons in an atom can vary. The **mass number** or **nucleon number** is the total number of protons and neutrons in the nucleus.

We can represent the atomic number and mass number of an element in symbol form, as shown in Figure C.

1 What does the atomic number of an element tell you?

2 What does the mass number of an element tell you?

3 A helium nucleus has 2 protons and 2 neutrons. Write down the symbol for the helium nucleus, showing the atomic number and mass number.

4 The symbol for a beryllium nucleus is $^9_4$Be. How many of the following particles does it contain?
a protons
b neutrons

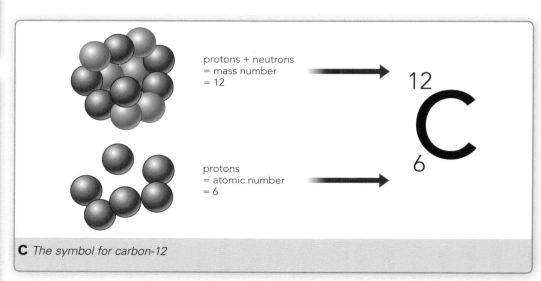

protons + neutrons = mass number = 12

protons = atomic number = 6

$^{12}_{6}$C

C The symbol for carbon-12

For example, an atom of sodium (Na) has 11 protons and 12 neutrons. Its mass number is 11 + 12 = 23. We can show its nucleus as $^{23}_{11}$Na.

252

Isotopes

Two atoms of the same element will always have the same atomic number. However, they can have different mass numbers if they contain different numbers of neutrons. Atoms of a single element that have different numbers of neutrons are called **isotopes**.

For example, atoms of lithium can exist as both lithium-6 and lithium-7. A number attached to an element's name shows we are referring to a particular isotope. The number is the mass number of the isotope. The nucleus of an atom of lithium-6 has three protons and three neutrons, so its mass number is $3 + 3 = 6$. The nucleus of an atom of lithium-7 has three protons and four neutrons, so its mass number is $3 + 4 = 7$.

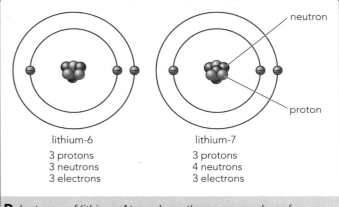

lithium-6	lithium-7
3 protons	3 protons
3 neutrons	4 neutrons
3 electrons	3 electrons

D *Isotopes of lithium. Atoms have the same number of electrons as protons, so these isotopes both have 3 electrons.*

5 For each of the nuclei shown in Figure E:
a work out its atomic number b look up its name using the periodic table in Appendix A. (The periodic table does not show the information in the same form as is used on this page, so take care when identifying the atomic number in the table.)
c write out its symbol (in the form show in Figure C).

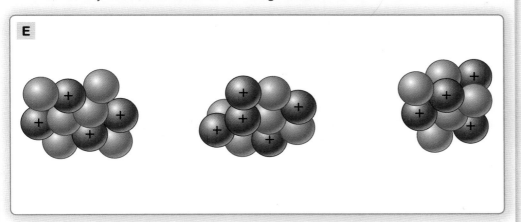

E

6 An atom of fluorine has 9 protons and 10 neutrons. Write down the symbol for a fluorine nucleus, showing the atomic number and nucleon number.

7 Sam says that as two different atoms have the same mass number, they must be isotopes. Is Sam correct? Explain your answer.

8 Hydrogen has three isotopes, hydrogen-1, hydrogen-2 and hydrogen-3. Explain what an isotope is and the similarities and differences between these three isotopes.

Skills spotlight

Scientists often use shortened ways of writing things down. Suggest why it is important that all scientists agree on the same symbols and conventions.

ResultsPlus
Watch Out!

Students often get the mass numbers and atomic numbers muddled in exams. Remember that the *mass* number is *more* than the atomic number.

Learning Outcomes

5.1 Describe the structure of nuclei of isotopes using the terms atomic (proton) number and mass (nucleon) number and using symbols in the format $_{6}^{14}C$

HSW 11 Present information using scientific conventions and symbols

What makes some isotopes radioactive?

The spectacular glow of the aurora borealis is caused by charged particles or 'cosmic rays' being funnelled towards the north pole by the Earth's magnetic field. Scientists have recently found this happens on other planets too.

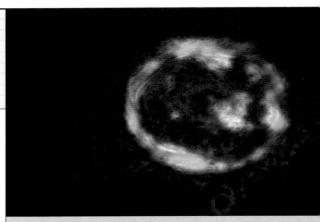

A *The aurora (in blue) over the north pole of Saturn, taken by the Cassini orbiter*

---

1 What is an ion?

2 A fluorine atom has seven electrons. It gains an electron when it becomes an ion. What charge does a fluoride ion have?

Protons carry a charge of +1. Neutrons have no charge and electrons carry a charge of -1. Since the number of electrons in an atom matches the number of protons, an atom has no overall charge. If an atom loses (or gains) an electron, it gains a charge and becomes an **ion**.

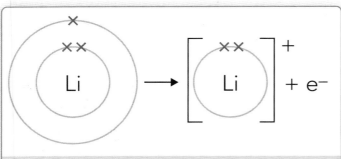

B *A lithium atom can lose one electron to become an ion. The brackets show that the whole ion is positively charged.*

Ionising radiation

Ionising radiation is radiation that has enough energy to cause atoms to lose electrons and become ions.

A **radioactive** substance has an **unstable** nucleus. The unstable nucleus **decays** to become more stable by losing energy when certain types of ionising radiation are emitted from the nucleus. These types are alpha particles, beta particles and gamma rays. You cannot predict when a nucleus will decay – it is a **random** process.

ResultsPlus
Watch Out!

Alpha, beta and gamma are not the only types of ionising radiation, but they are the commonest types of ionising radiation produced by unstable nuclei.

3 What is the mass number of an alpha particle?

Alpha particles contain two protons and two neutrons, just like the nucleus of a helium atom. They have no electrons and so have a charge of +2.

Alpha particles are emitted from the nucleus at high speeds. However, each time they ionise an atom they lose some energy. Since they produce many ions in a short distance, they lose energy quickly and have a short **penetration distance**. They are stopped by a few centimetres of air or a few millimetres of paper.

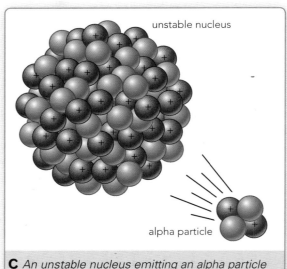

unstable nucleus

alpha particle

C *An unstable nucleus emitting an alpha particle*

Beta particles are electrons that are emitted from an unstable nucleus. They are much less ionising than alpha particles and so can penetrate much further into matter than alpha particles can. They can be stopped by a few millimetres of aluminium or by an even smaller thickness of lead.

Gamma rays are high-frequency electromagnetic waves emitted by some unstable nuclei and so travel at the speed of light. They do not have an electric charge.

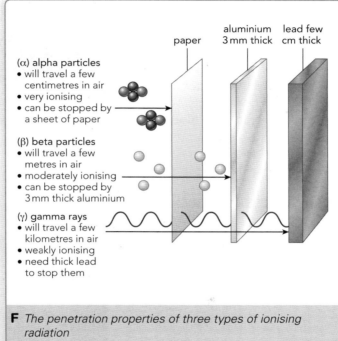

D *An unstable nucleus emitting a beta particle*

Gamma rays are about ten times less ionising than beta particles and can penetrate matter very easily. They can be partly stopped by a few centimetres of lead and absorbed by many metres of concrete.

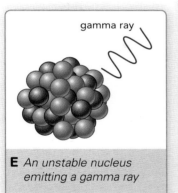

E *An unstable nucleus emitting a gamma ray*

(α) alpha particles
- will travel a few centimetres in air
- very ionising
- can be stopped by a sheet of paper

(β) beta particles
- will travel a few metres in air
- moderately ionising
- can be stopped by 3 mm thick aluminium

(γ) gamma rays
- will travel a few kilometres in air
- weakly ionising
- need thick lead to stop them

paper   aluminium 3 mm thick   lead few cm thick

F *The penetration properties of three types of ionising radiation*

4 What materials will absorb and stop beta particles?

5 The reactor in a nuclear power station is surrounded by large amounts of concrete. Why is this necessary?

Skills spotlight

Scientific ideas can explain how things work. A 'Geiger counter' counts the number of ions produced by ionising radiation inside its tube. Explain how this instrument can be used to identify the type of radiation emitted by a radioactive source.

6 Draw up a table to summarise the form, charge, penetration and ionisation properties of the three types of radiation.

7 What changes take place in a nucleus when it emits an alpha particle?

8 Explain how an oxygen molecule in the air might become an ion by being near a radioactive source.

Learning Outcomes

5.2 Explain how atoms may gain or lose electrons to form ions

5.3 Recall that alpha and beta particles and gamma rays are ionising radiations emitted from unstable nuclei in a random process

5.4 Recall that an alpha particle is equivalent to a helium nucleus, a beta particle is an electron emitted from the nucleus and a gamma ray is electromagnetic radiation

5.5 Compare alpha, beta and gamma radiations in terms of their abilities to penetrate and ionise

HSW **3** Describe how phenomena are explained using scientific theories and ideas

How do nuclear reactions provide an energy source?

The Cassini-Huygens space probe, launched in 1997, is now orbiting Saturn. The electrical power needed for its scientific instruments and communications equipment is generated from the thermal energy produced by radioactive plutonium as it decays.

Watch Out!

Students often get nuclear reactions and chemical reactions mixed up in exams – they are not the same thing. In a chemical reaction, the atoms themselves do not change. In a nuclear reaction such as fission, the nuclei of the atoms change.

1 What particle triggers nuclear fission in uranium-235?

2 What are the products of the fission of uranium-235 in the example in Figure B?

Radioactive decay

The process of radioactive decay releases energy. Alpha and beta particles are emitted at high speed and so have kinetic energy. Gamma rays have energy in the form of electromagnetic radiation.

Other nuclear reactions

Other types of nuclear reaction can also release energy. In some large unstable nuclei, the nucleus can split into two smaller nuclei called **daughter nuclei**. This process is called **nuclear fission**. For example, when a uranium-235 nucleus absorbs a neutron it immediately splits into two smaller daughter nuclei. Two or more neutrons are also released. This is not the same as radioactive decay, when an alpha or beta particle or gamma radiation is emitted from the nucleus.

A The Cassini orbiter uses radioactivity as a power source.

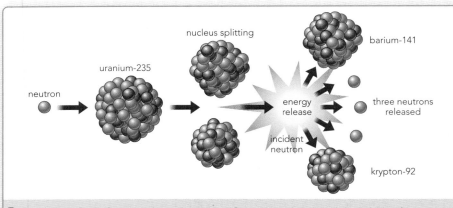

B Example of the fission of uranium-235. Other daughter nuclei could be formed.

Nuclear fission releases a huge amount of energy – 1 g of uranium releases 10 000 times more energy than burning 1 g of oil.

Most of the energy is in the form of kinetic energy. Both daughter nuclei and the neutrons have a lot of kinetic energy because they are moving at high speeds. Thermal energy is also released.

If the neutrons released are absorbed by other uranium-235 nuclei, each of these nuclei will become unstable and release more neutrons when their nucleus splits. These neutrons can then be absorbed by yet more uranium nuclei, which in turn split up, releasing more neutrons. This is an uncontrolled nuclear **chain reaction**, such as occurs in an atomic bomb.

The chain reaction can be controlled if other materials absorb some of the neutrons. A chain reaction can be created in which only one of the neutrons produced by fission is absorbed by another uranium-235 nucleus. This regulates the amount of energy produced and is the type of chain reaction that occurs in a nuclear reactor.

C *Enormous amounts of energy are released in a nuclear explosion.*

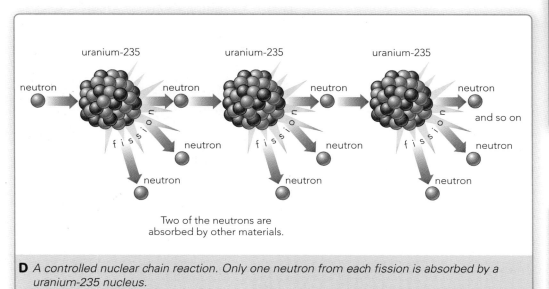

Two of the neutrons are absorbed by other materials.

D *A controlled nuclear chain reaction. Only one neutron from each fission is absorbed by a uranium-235 nucleus.*

3 What form of energy is released by nuclear fission?

4 How can a chain reaction be controlled?

Nuclear fusion

Instead of splitting large nuclei, small nuclei can be combined to form larger ones. The process is called **nuclear fusion** and it also releases lots of energy.

5 Why does an atomic bomb release a large amount of energy in a short time?

6 Match each of these sentences to one of the following: no chain reaction, controlled chain reaction, uncontrolled chain reaction.
a An average of more than one neutron per fission causes another fission.
b An average of less than one neutron per fission causes another fission.
c Exactly one neutron per fission causes another fission.

7 Explain the principle of a controlled nuclear chain reaction.

Skills spotlight

The number of fissions differs in controlled and uncontrolled chain reactions. Work out the number of atoms of uranium-235 that have split up after 10 cycles when one, two and three neutrons go on to trigger new fissions.

Learning Outcomes

5.6 Demonstrate an understanding that nuclear reactions can be a source of energy, including fission, fusion and radioactive decay

5.7 Explain how the fission of U-235 produces two daughter nuclei and two or more neutrons, accompanied by a release of energy

5.8 Explain the principle of a controlled nuclear chain reaction

5.13 Explain the difference between nuclear fusion and nuclear fission

HSW **10** Use qualitative and quantitative approaches when presenting scientific ideas and arguments, and recording observations

How does a nuclear reactor work?

In 1978 radioactive material was spread over a large area of Canada. The Russian satellite Kosmos 954, which was powered by a nuclear reactor, re-entered the Earth's atmosphere and broke up. Nuclear power plants are still used in spacecraft today but some people think this should be banned.

A *Radioactive debris from Kosmos 954*

1 What energy transfer takes place in a nuclear reactor? **?**

Nuclear reactors in nuclear power stations transform energy contained in the nuclei of uranium and plutonium atoms (nuclear energy) into thermal energy using nuclear fission. The mainly uranium fuel is made into pellets, which are inserted into hollow fuel rods several metres long. The rods are placed inside the reactor **core**.

The rate at which nuclear energy is transferred to thermal energy is kept constant by controlling the chain reaction. If just one of the neutrons released by decay of a uranium nucleus is absorbed by another uranium nucleus (causing it to decay), then the chain reaction continues at a constant rate. The number of free neutrons will not increase or decrease.

B *An array of fuel rods*

To control the fission chain reaction, the extra neutrons released by the decay of each nucleus have to be absorbed. The **control rods** contain elements that absorb neutrons. These rods are placed between the fuel rods in the reactor core.

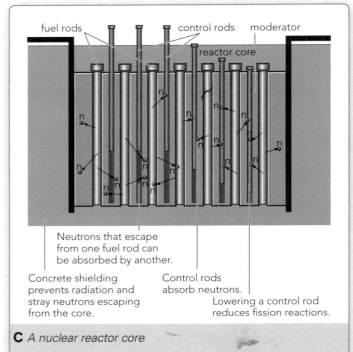

Neutrons that escape from one fuel rod can be absorbed by another.

Concrete shielding prevents radiation and stray neutrons escaping from the core.

Control rods absorb neutrons.

Lowering a control rod reduces fission reactions.

C *A nuclear reactor core*

2 Why must the chain reaction in a nuclear reactor be controlled? **?**

If the rate of fission needs to be increased, the control rods are moved out of the core so that fewer neutrons are absorbed, and vice versa. When the control rods are fully lowered into the core, the chain reaction stops and the reactor shuts down.

Moderators

The neutrons emitted from the fission of a uranium-235 nucleus are moving very fast. They need to be slowed down so that it is more likely that they will be absorbed by another uranium-235 nucleus. Another material in the core, called a **moderator**, slows down the neutrons.

Generating electricity

Thermal energy from the core is transferred to a coolant, which is pumped through the reactor. The coolant is usually water at high pressure. This super-heated water is pumped to a 'heat exchanger' where it is used to produce steam. The steam drives a turbine, which turns a generator. The generator transfers kinetic energy to electrical energy.

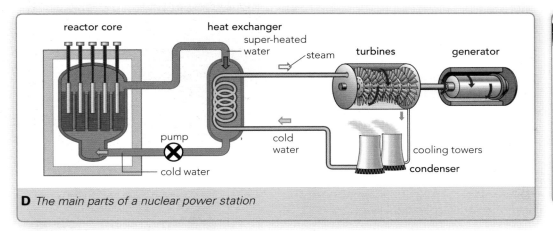

D The main parts of a nuclear power station

Radioactive waste

After a period of time, **radioactive waste** builds up in the reactor core. This waste is made up of radioactive daughter nuclei and radioactive isotopes formed when the materials in the core absorb neutrons.

4 What precautions would you need to take when storing fuel rods?

5 Draw a flow diagram to show how nuclear energy is transformed into electrical energy in a nuclear power station.

6 Explain what is done to increase the temperature of the core of a nuclear reactor.

7 What materials in a reactor core can affect the speed of the neutrons emitted in nuclear fission?

8 Describe how control rods are used to control the rate of the chain reaction in a reactor core.

---

3 In a nuclear reactor, what are the functions of:
a the control rods
b a moderator?

Skills spotlight

There are social, environmental and economic factors involved in the decision to build a nuclear power station. What issues need to be considered when deciding if and where a nuclear power station should be built?

Results Plus
Watch Out!

In exams students often lose marks by stating that the rate of a nuclear reaction can be changed by changing the moderator. The control rods can be lowered to absorb more neutrons and act as the moderator, to slow down the reaction.

Learning Outcomes

5.9 Explain how the chain reaction is controlled in a nuclear reactor including the action of moderators and control rods

5.10 Describe how thermal (heat) energy from the chain reaction is converted into electrical energy in a nuclear power station

5.11 Recall that the products of nuclear fission are radioactive

HSW **13** Understand how and why decisions about science and technology are made, including those that raise ethical issues, and about the social, economic and environmental effects of such decisions

What is nuclear fusion?

In 1958, Ford developed a concept car called the nucleon, which was to be powered by a small nuclear reactor. However, safety concerns mean that such a car will only ever exist in films.

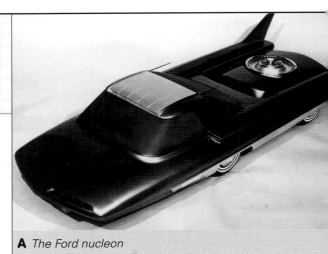

A *The Ford nucleon*

Nuclear fusion occurs when small nuclei combine to form larger ones. A fusion reaction in which hydrogen nuclei combine to form helium is the energy source for stars, including our Sun. Scientists are investigating using a similar fusion reaction here on Earth to generate electricity. This reaction fuses two isotopes of hydrogen, hydrogen-2 (deuterium) and hydrogen-3 (tritium), to form helium. The reaction frees a neutron from the nucleus and releases an enormous amount of energy.

The helium produced in nuclear fusion is not radioactive but any materials used to contain fusion do become radioactive.

Getting new ideas accepted

New scientific theories have to be **validated** by the scientific community before they are accepted. A report of the method and results is **peer-reviewed** – different scientists working in the same field check it. Other scientists must be able to carry out the experiment and get the same results.

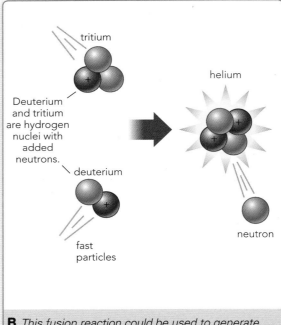

B *This fusion reaction could be used to generate electricity from the energy carried by the neutrons.*

Fusion and temperature

In nuclear fusion research, a deuterium nucleus and a tritium nucleus have to be forced together. The nuclei are positively charged and like charges repel. This is called **electrostatic repulsion**.

---

1 What is the source of energy in the Sun?

2 What are the products of the fusion of hydrogen?

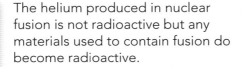

Skills spotlight

In 1989, Martin Fleischmann and Stanley Pons claimed to have achieved nuclear fusion at 50 °C, raising hopes of a cheap and carbon-free source of energy. However, their results have not been accepted. Suggest what criteria scientists used to reject their claim.

For the nuclei to fuse, they need to get within a million millionth of a millimetre of each other. The Sun has a very strong gravitational field, which creates very high densities of nuclei at its centre. Nuclei are much more likely to collide if their density is higher. We can't create these conditions on Earth but fusion reactors try to produce very high pressures.

If the nuclei are travelling fast enough, some can overcome their electrostatic repulsion and collide. The higher the temperature, the faster the nuclei move, and the more likely they are to collide. This means that the temperature inside a fusion reactor must be very high – about 150 000 000 °C (ten times hotter than the Sun).

It is very difficult to sustain the very high temperature and pressure required for fusion. So far none of the experimental reactors have produced more energy than has been put in, so fusion power is a long way off.

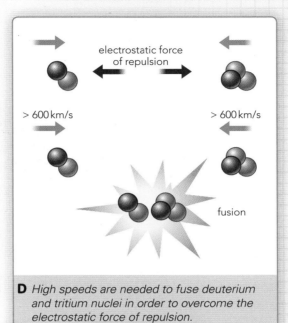

D High speeds are needed to fuse deuterium and tritium nuclei in order to overcome the electrostatic force of repulsion.

C This fusion reactor uses 192 lasers, all pointed at a 2 mm wide target containing tritium and deuterium gases at its centre.

H 3 Why does fusion not happen at room temperature?

H 4 Why does fusion happen at a lower temperature in the Sun than in a fusion reactor on Earth?

5 Explain the differences between nuclear fission and nuclear fusion.

H 6 Use scientific ideas to suggest why the idea of 'cold fusion' from Pons and Fleischmann's results was not accepted.

7 Describe the difficulties of building a commercial fusion power station.

Learning Outcomes

5.12 Describe nuclear fusion as the creation of larger nuclei from smaller nuclei, accompanied by a release of energy and recognise fusion as the energy source for stars

5.13 Explain the difference between nuclear fusion and nuclear fission

H 5.14 Explain why nuclear fusion does not happen at low temperatures and pressures, due to electrostatic repulsion of protons

H 5.15 Relate the conditions for fusion to the difficulty of making a practical and economic form of power station

5.16 Demonstrate an understanding that new scientific theories, such as 'cold fusion', are not accepted until they have been validated by the scientific community

HSW 14 Describe how scientists share data and discuss new ideas, and how over time this process helps to reduce uncertainties and revise scientific theories

How have ideas about radioactivity changed?

Marie Curie (1867–1934) worked extensively with radioactive materials. Her laboratory notebooks are now considered too dangerous to handle because of the radioactive fingerprints. Even her cookbook is highly radioactive. These papers are now kept in lead-lined boxes – if researchers want to view them, they have to wear protective clothing.

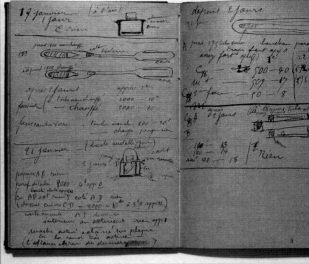

A One of Marie Curie's notebooks

Henri Becquerel (1852–1908) investigated how uranium phosphoresces (emits light) after being exposed to sunlight. He observed that 'invisible rays' also emitted by uranium could expose (darken) photographic plates. In 1896 during some cloudy weather, he put some uranium away in a drawer with a photographic plate. The plate unexpectedly became exposed. Becquerel had discovered that the radiation emitted by uranium did not need an external energy source. He had discovered radioactivity.

Becquerel then showed that this radiation could ionise gases. He noticed his skin was burned after accidentally leaving a container of radium in his pocket but he did not realise that ionisation could occur in cells and was the cause of the damage.

Marie Curie and her husband Pierre (1859–1906) also burned their hands by handling radioactive materials such as radium. The Curies discovered that some materials emitted much more radiation than others.

1 a What did Becquerel discover?
b What did the Curies discover about radioactivity?

Changing ideas about safety

In the 1920s scientists began to link cancer and other health problems to the effects of ionising radiation. Marie Curie and her daughter Irene both died from leukaemia, which was probably caused by the radioactive materials they worked with.

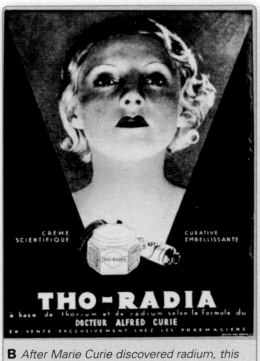

B After Marie Curie discovered radium, this element became so popular it was soon put in food and even water. This face cream, advertised in 1933, contained radioactive thorium and radium.

We now know that a large amount of ionising radiation can cause tissue damage such as reddened skin (radiation burns) and also other effects that cannot be seen. Much smaller amounts of ionising radiation over long periods of time can damage the DNA inside a cell. This damage is called a **mutation**. DNA contains the instructions controlling a cell, so some mutations can cause the cell to malfunction and may cause cancer.

Handling radioactive sources

We now know about the **hazards** of radioactive sources and how to protect people who work with radioactivity. The **risk** of harm decreases with distance from the source, so sources are always handled with tongs. The risk can also be reduced by not pointing sources at people. We also know that all but the most penetrating radiation is stopped by a few millimetres of lead, so sources are kept in a lead-lined container.

2 a What does ionising radiation do to cells?
b What can this lead to?

3 Suggest why experts in the early 20th century thought that working with small amounts of radioactive materials was safe.

C *Protective clothing worn by a radiation worker*

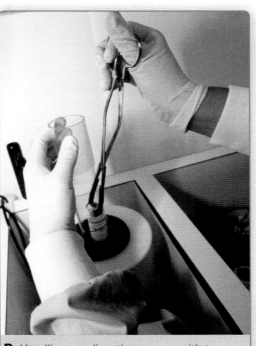

D *Handling a radioactive source with tongs*

Skills spotlight

Scientific knowledge changes over time as more observations and data are collected. Which observations about the effects of radioactivity did Marie and Pierre Curie know about? What effects did they not know about?

4 What caused ideas about the safety of ionising radiation to change in the 1920s?

5 a List four precautions that someone working with radioactivity today needs to take to reduce the risks from radiation.
b Explain why each precaution should be taken.

ResultsPlus
Watch Out!

Not all forms of radiation are ionising. Light and microwaves are not ionising radiation as they have low frequency and low energy. They do not have enough energy to ionise atoms.

Learning Outcomes

6.9 Demonstrate an understanding of the dangers of ionising radiation in terms of tissue damage and possible mutations and relate this to the precautions needed

6.10 Describe how scientists have changed their ideas of radioactivity over time, including:
a the awareness of the hazards associated with radioactive sources
b why the scientific ideas change over time

HSW 14 Describe how scientists share data and discuss new ideas, and how over time this process helps to reduce uncertainties and revise scientific theories

What happens to nuclear waste?

From 1953 until the late 1980s, British nuclear waste was dumped at sea. Over 28 000 barrels of waste were dumped off the Channel Islands. These barrels are now corroding and breaking up and could be releasing radioactive materials into the sea.

A *A corroding barrel of waste on the sea bed near the Channel Islands*

Types of waste

The fission products from the uranium fuel used in nuclear power stations are very **radioactive**. This is **high level waste** (**HLW**) and it produces large amounts of ionising radiation for about 50 years.

1 Where does HLW come from?

After about 50 years HLW is much less radioactive but it remains moderately radioactive for many tens of thousands of years. It is now **intermediate level waste** (**ILW**). ILW also includes the metal cylinders that contained the uranium fuel. These were changed by radiation from the fuel and became radioactive themselves.

Low level waste (**LLW**) is only slightly radioactive but will remain so for tens of thousands of years, like ILW. It includes clothing and cleaning materials from nuclear power stations. Hospitals are also a source of LLW because of the radioactive isotopes used in radiotherapy treatment for cancer.

2 What are the similarities between ILW and LLW?

3 What are the differences between HLW and ILW?

4 How is HLW stored?

Storage and disposal of nuclear waste

HLW is transported inside thick concrete and steel containers, which absorb the radiation. It is then sealed in glass to stop the radioactive material escaping, and stored in canisters until the waste is less radioactive (and has become ILW).

ILW is currently stored in concrete and steel containers, but none has been disposed of yet.

LLW is compacted and buried in special landfill sites because of the possibility of radioactive material leaking into soil or water.

B *HLW from all UK nuclear power stations is transported by train to Sellafield in Cumbria.*

ResultsPlus
Watch Out!

Many students lose marks by saying that nuclear power stations make the local area more radioactive. Remember that radioactive waste is only released into the environment if there is an accident.

Method of disposal	Problems
Firing into space	Launch vehicle could fall back to Earth, spreading radioactive material over a wide area.
Dumping in barrels at sea	Barrels can corrode and release radioactive materials which could enter the food chain.
Storage underground	Site needs to be geologically stable, i.e. very low risk of earthquakes.

C *Some of the options for disposing of nuclear waste*

Advantages and disadvantages of nuclear power

A nuclear power station itself does not produce carbon dioxide (which contributes to global warming). However, the processes used to make the fuel rods require energy and generating this energy may produce carbon dioxide.

Nuclear waste has to be stored for tens of thousands of years, until the radioactivity has decreased to safer levels. During this time, nothing should leak from the waste into the environment.

In 1986, the reactor at the Chernobyl nuclear power station in the Ukraine exploded. Radioactive material was spread by wind over a very wide area before falling out with the rain, including over the UK. Some people think that nuclear power is unsafe because of the risk of accidents like the one at Chernobyl.

D *Anatoly Rasskazov flew over Chernobyl the day after the accident to take official photos. The radiation burn on his forehead is still there.*

5 The UK government has decided that ILW should be buried deep underground. Suggest some factors that could affect the choice of a long-term deep disposal site.

Skills spotlight

In a scientific discussion, evidence is presented for and against an idea and then used to reach a decision for or against. Write an argument in support of one method of disposing of nuclear waste.

6 a Describe the advantages of nuclear power.
b Explain why some people think that nuclear power is unsafe.

7 Explain why it is difficult to dispose of radioactive waste safely.

6.11 Discuss the long-term possibilities for storage and disposal of nuclear waste

6.12 Evaluate the advantages and disadvantages of nuclear power for generating electricity, including the lack of carbon dioxide emissions, risks, public perception, waste disposal and safety issues

HSW 11 Present information, develop an argument and draw a conclusion, using scientific, technical and mathematical language

What is the half-life of a radioactive material?

After Chernobyl, Arctic lichens absorbed large amounts of radioactive caesium-137. The lichens are eaten by reindeer, which are eaten by humans. Some reindeer herders in the Arctic have caesium-137 levels that are 300 times higher than normal. This situation will continue for hundreds of years because caesium-137 decays slowly.

A

When an unstable nucleus undergoes **radioactive decay** its nucleus changes to become more stable. The **activity** of any radioactive substance is the number of nuclear decays per second and is measured in **becquerel (Bq)**. 1 Bq is one nuclear decay each second.

Radioactive decay is a random process – we cannot predict when it will happen. When you throw a die, sometimes you get a six and sometimes you don't. In a similar way, at any given moment there is a certain probability that a particular unstable nucleus will decay.

The **half-life** is the time taken for half the unstable nuclei in a sample of a radioactive isotope to decay. Some isotopes have half-lives of fractions of a second but others have half-lives of millions of years. The half-life is the same for however much of an isotope there is.

1 How many nuclei decay each second in a sample of uranium-235 that has an activity of 2 Bq?

2 What is the half-life of an isotope?

3 A 10 kg sample of caesium-137 has a half-life of 30 years. What will the half-life of a 5 kg sample be?

4 Strontium-90 has a half-life of 29 years. How many strontium-90 half-lives is:
a 29 years
b 58 years
c 116 years
d 14.5 years?

5 There are 10 million atoms in a sample of radon-222. How many undecayed nuclei are left after:
a 3.8 days
b 7.6 days
c 11.4 days?

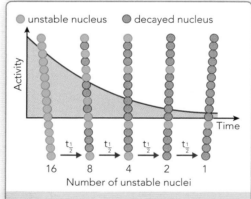

B After each half-life, the number of unstable nuclei halves.

Isotope	Half-life	Ionising radiation emitted
uranium-235	700 million years	alpha particle
technetium-99	211 000 years	beta particle
carbon-14	5730 years	beta particle
strontium-90	29 years	beta particle
caesium-137	30 years	beta particle
radon-222	3.8 days	alpha particle
thallium-208	3.1 minutes	beta particle
polonium-216	0.15 seconds	alpha particle

C Half-lives of some radioactive isotopes

After decaying, a nucleus may become stable. The more stable nuclei a sample of a substance contains, the lower its activity. The half-life of an isotope is found by recording the activity of a sample over a period of time.

The radioactivity of a source is measured using a **Geiger-Müller (GM) tube**. A GM tube can be connected to a counter or alternatively it may give a click each time ionising radiation is detected in the GM tube. The **count rate** is the number of clicks per second or minute. You can work out how long it takes the count rate to halve from a graph.

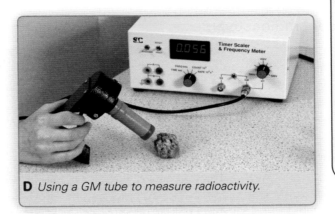

D Using a GM tube to measure radioactivity.

In Figure E, the count rate at 5 minutes is 600 counts per second. After one half-life the count rate will have decreased to 300 counts per second. This occurs at 12 minutes. So the half-life is 12 minutes – 5 minutes = 7 minutes.

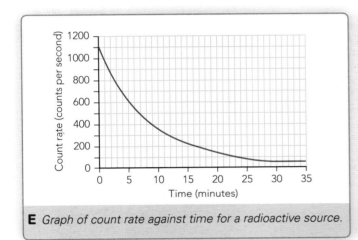

E Graph of count rate against time for a radioactive source.

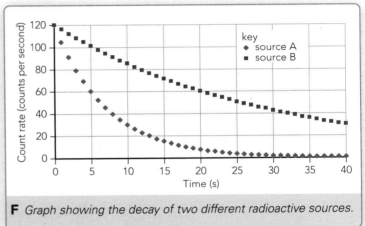

F Graph showing the decay of two different radioactive sources.

6 Work out the half-lives of the two sources shown in Figure F.

7 Explain how the activity of a radioactive sample relates to the number of unstable nuclei.

Skills spotlight

Scientists often collect quantitative data to answer a question. Describe how you could find the half-life of a newly discovered radioactive substance.

Learning Outcomes

6.4 Describe how the activity of a radioactive source decreases over a period of time

6.5 Recall that the unit of activity of a radioactive isotope is the becquerel, Bq

6.6 Recall that the half-life of a radioactive isotope is the time taken for half the undecayed nuclei to decay

6.7 Use the concept of half-life to carry out simple calculations on the decay of a radioactive isotope, including graphical representations

HSW 10 Use qualitative and quantitative approaches when presenting scientific ideas and arguments, and recording observations

Is the landing of dropped buttered toast a random process?

P2.31 Investigating radioactive decay models

How can you model radioactive decay?

In Sweden each December Nobel Prizes are awarded to scientists who have made big contributions to science. In the USA each October, 'Ig® Nobel Prizes' are given to scientists whose research makes people laugh and then makes them think. The Ig® Nobel physics prize for 2009 was won by Prof. Robert Matthews from Aston University, Birmingham, for his research showing that the falling of buttered toast is not a random process – and that toast does land buttered side down more often than not.

A *The Ig® Nobel Awards are given out by a real Nobel Prize winner.*

Radioactive decay is a random process. So a model for radioactive decay needs to be based on something that is also random. Flipping a coin is a random process, so you can use it to model radioactive decay.

You can think of an unstable nucleus as having two states – undecayed and decayed. A coin also has two states when you flip it – heads and tails.

If you flip a large number of coins once, you would expect that about half of them would land showing heads and the other half tails. This can be used as a model of a half-life – half of the coins are in each 'state'. Rolling dice is also a random process and can be used to model radioactive decay.

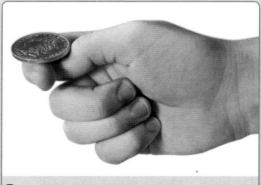

B *Using coins to model radioactive decay.*

Your task

You are going to plan an investigation that will allow you to find out about a model that simulates radioactive decay. Your teacher will provide you with some materials to help you organise this task.

Learning Outcomes

6.8 Investigate models which simulate radioactive decay

Build Better Answers

When planning an investigation like this, one of the skills you will be assessed on is your ability to *evaluate your conclusion*. There are 4 marks available for this skill. Here are two extracts focusing on this skill. Other skills are dealt with in other lessons.

Student extract 1 — A basic response for this skill

The student has not explained *how* it matches.

The student does suggest how to strengthen the conclusion.

> The pattern I obtained matches that of radioactive decay. My conclusion would have been supported more if I had used even more cubes to start off with.

This is correct but it is not linked to the theory and does not say if the conclusion is a good one or not.

Student extract 2 — A good response for this skill

The hypothesis has been linked in with scientific ideas.

The secondary evidence has also been linked to the conclusion.

> My conclusion is strong, but the number of cubes you start off with could be increased to provide stronger evidence. My conclusion supports my hypothesis and matches my scientific knowledge. As radioactive decay is a random process and there are large numbers of nuclei in even a small sample, larger numbers of cubes are needed to model radioactive decay more effectively. This is backed up by the simulation I found on the internet.

An explanation is given of how the investigation could be improved. The student has also included ways that the investigation could be extended.

ResultsPlus

To access 2 marks

- Evaluate how well all your evidence supports your conclusion
- Suggest how your evidence can be improved to strengthen your conclusion

To access 4 marks

You also need to:
- Evaluate how well other scientific ideas support your conclusion
- Suggest how the investigation could be extended to support your conclusion

What is background radiation?

The nuclear facility at Sellafield in Cumbria is permitted to release a certain amount of radioactive material into the air or sea. It is allowed to do this because the activity of this material is so low that it does not contribute significantly to natural levels of radioactivity.

A *Part of the nuclear facility in Sellafield, Cumbria*

> **1** What is background radiation?
>
> **2** Tom records background counts of 15, 22 and 17 counts per minute. He then records a count rate of 186 counts per minute from a sample of granite. What is the corrected count for the sample's activity?

We are constantly being exposed to ionising radiation at a low level, from space and from naturally radioactive substances in the environment. This is called **background radiation**.

When scientists measure the activity of a source, they need to measure the **background count** first by taking several readings and averaging them. This background count is then subtracted from measurements to give a corrected reading of the source's activity.

Sources of background radiation

Figure B shows the sources of background radiation averaged over the UK. The main source is **radon** gas. When uranium in rocks decays, it produces other radioactive isotopes that also decay. One of these is a gas called radon.

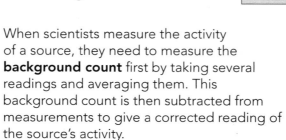

B *Sources of background radiation in the UK*

> **3** Where does radon gas come from?
>
> **4** Why do you think radon gas could be dangerous?

Radon diffuses into the air from rocks and soil and can build up in houses, especially where there is poor ventilation. Radon has a half-life of 3.8 days and decays by giving out an alpha particle. The amount of radon in the air depends on the type of rock and its uranium content. Rock type and building stone vary around the country and so does the amount of radon.

C *This radon outlet pipe sucks air containing radon from beneath a solid concrete floor, stopping it from entering the house.*

The contribution from medical sources results from having X-rays and gamma ray scans, and from treatments for cancer. Some foods naturally contain small amounts of radioactive substances.

High-energy charged particles stream out of the Sun and other stars, as well as supernovae, neutron stars and black holes. They are known as **cosmic rays** and are a form of ionising radiation. Many cosmic rays are stopped by the upper atmosphere but some still reach the Earth's surface.

D *A cosmic ray detector has been put beneath the Pyramid of the Sun in Teotihuacan, Mexico. By detecting whether cosmic rays are stopped more by some parts of the pyramid than others, scientists hope to find hidden chambers.*

7 Explain why background radiation varies in different parts of the UK. ?

8 The following was found on a blog: 'Natural radiation won't hurt you but human-made radiation will.' Write a response to this blogger.

5 How do you think your food and drink becomes naturally radioactive? ?

6 During a 'solar storm' the Sun's output of high-energy charged particles increases dramatically. Explain why some scientists suggest that aeroplanes should fly at lower altitudes during a solar storm. ?

Skills spotlight

Scientists collect and analyse data on background radiation. What data would you need to find to compare how people's exposure to background radiation varies in the UK?

ResultsPlus
Watch Out!

Many students lose marks in cosmic ray questions because they do not state that cosmic rays are high energy particles that can ionise atoms. Another type of ionising radiation.

Learning Outcomes

6.1 Explain what is meant by background radiation, including how regional variations within the UK are caused in particular by radon gas

6.2 Recall the origins of background radiation from Earth and space

HSW *1* Explain how scientific data is collected and analysed

How is ionising radiation used in medicine and in food preparation?

Marie Curie worked with doctors to develop a treatment using radium that could reduce or even cure cancer tumours. She collected radon gas and sealed it in thin glass tubes about 1 cm long. These tubes were then covered in platinum and placed in a person's body. Cervical cancer was first treated in this way in 1905.

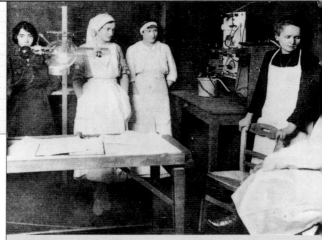

A *Marie Curie explains the potential benefits of radium treatment to nurses.*

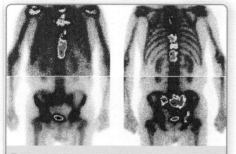

B *A radioactive tracer shows bone tumours in this gamma ray scan. The areas of bone cancer are coloured green.*

Diagnosis of cancer

Gamma rays can be used to help diagnose cancer. A **tracer** solution containing a source of gamma rays is injected into the body. The tracer collects where there are areas of abnormal activity, e.g. cancers. The gamma rays are then detected using a 'gamma camera'.

Gamma sources are used because they pass straight through the body, allowing them to be detected.

Treatment of cancer

Ionising radiation can also be used to treat diseases such as cancer. This is known as **radiotherapy**. Some ways of doing this involve radioactive substances that produce gamma rays. For example, gamma rays can be used as beams of radiation to target and kill cancer cells. Radioactive sources can also be put inside a person for a limited amount of time at the place where the cancer is.

1 Why are alpha and beta sources not used in radioactive tracers to diagnose cancer?

2 What is radiotherapy?

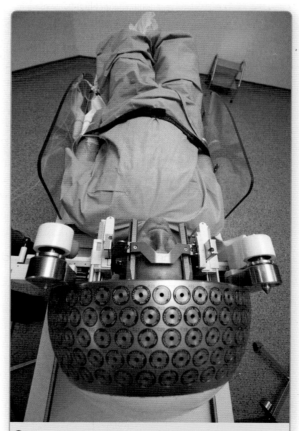

C *This patient is about to have treatment with a 'gamma knife'. The patient's head is held in place so that it cannot move while the gamma rays are aimed.*

Sterilisation of equipment

Before they can be used on patients, surgical instruments need to be **sterilised** to kill microorganisms. The usual method is to heat them. Some instruments such as plastic syringes cannot be sterilised using heat, so they are **irradiated** with gamma rays.

Irradiating food

All food contains bacteria in varying amounts. The bacteria eventually cause the food to decompose. Some types of bacteria also cause food poisoning.

Food can be irradiated with gamma rays to kill bacteria. This makes it safer to eat and also means that it can be stored for longer before going off. Irradiating the food also kills any pests such as insects that may be in it. It does not make the food more radioactive, although some foods are naturally radioactive.

Types of foods that are allowed to be irradiated in the UK are fruit, vegetables, cereals, bulbs (e.g. onions) and tubers (e.g. potatoes), herbs and spices, fish, shellfish and poultry.

D *These containers are full of medical equipment and are being positioned over the source of gamma radiation.*

3 a Why does surgical equipment need to be sterilised?
b Why is some of the equipment sterilised by irradiation?

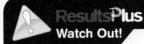

Watch Out!

Don't get irradiated and radioactive mixed up. Irradiated things are not radioactive – they have just been exposed to ionising radiation.

Skills spotlight

In the UK, irradiated food must be labelled with the words 'irradiated' or 'treated with ionising radiation'. Discuss the advantages and disadvantages of labelling food that has been irradiated. Try to include any ethical issues about using these technologies.

4 What are the benefits and drawbacks of irradiating food?

5 Explain what happens when food is irradiated.

6 Explain what the properties of a radioactive tracer used to diagnose medical problems should be.

Learning Outcomes

6.3 Describe uses of radioactivity, including:
b irradiating food
c sterilisation of equipment
e diagnosis and treatment of cancer

HSW 13a Explain how and why decisions about uses of science and technology are made

>>>>>>>>>>>>>>>>>>>>>>> Where can you find a source of radioactive americium in a house?

What are some of the other uses of ionising radiation?

In 1994, a 17-year-old boy called David Hahn tried to build a nuclear reactor in a garden shed using over 100 smoke detectors. His Geiger counter detected increased levels of radioactivity from down the street so he tried to dispose of his project.

Smoke alarms

A smoke alarm contains a source of alpha particles, usually a radioisotope called americium-241. There is an electrical circuit with an air gap between two electrically charged plates. The americium-241 source releases alpha particles, which ionise the air to give electrons (knocked out of the air molecules) and ions. These charged particles are attracted to opposite charged plates and so allow a small electrical current to flow.

1 What type of radiation is produced by americium-241? ?

A David Hahn was arrested and his 'project' was placed into 39 barrels like these ones, sealed and buried at a radioactive disposal site.

The air is constantly being ionised by the americium-241 source. This means that there is a constant electric current. As long as this current is flowing, the alarm will not sound.

When smoke gets into the air gap, the smoke particles absorb the alpha particles. This means that the current flowing across the gap decreases. The alarm sounds when the current drops below a certain level.

Checking thicknesses

Paper is made by squeezing wood pulp between rollers. Paper can be made in different thicknesses and the rollers need to squeeze the wood pulp with a force that produces the correct thickness of paper.

The detector in Figure C counts the rate at which beta particles get through the paper from a source on one side.

Smoke enters smoke detector.

Alpha particles ionise the air and these charged particles move across the gap forming a current.

Smoke in the machine will absorb alpha particles so make the current fall.

Americium-241 alpha source

Siren will sound when the current falls.

detector

Americium-241 source gives off a constant stream of alpha particles.

battery

A detector senses the amount of current.

B How a smoke detector works

When the paper is too thin, more beta particles penetrate the paper and the detector records a higher count rate. A computer senses that the count rate has risen and reduces the pressure applied to the rollers to make the paper thicker. When the paper is too thick, the opposite happens.

Tracers in the environment

Radioactive isotopes can also be added to water and used as tracers, for example to monitor pollution. A gamma source added to water is used to detect leaks in pipes buried underground, such as water pipes. Where there is a leak, water leaks into the surrounding earth. A GM tube following the path of the pipe will detect higher levels of radiation where there is a leak.

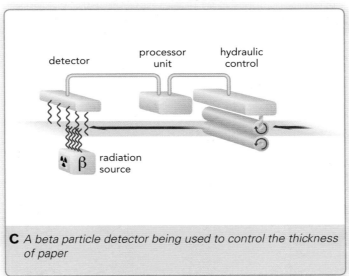

C A beta particle detector being used to control the thickness of paper

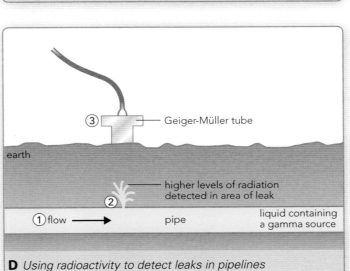

D Using radioactivity to detect leaks in pipelines

2 Look at Figure C. What happens when the paper is too thick?

3 Why would you not use an alpha source to monitor paper thickness?

Skills spotlight

There are always benefits, drawbacks and risks involved in using radioactive materials. How would you decide whether the benefits of smoke detectors outweigh the drawbacks and risks?

ResultsPlus
Watch Out!

Be careful to match examples of the uses of radioactivity correctly to the types of radiation. Many students lose marks because they get the properties of alpha and beta particles and gamma rays mixed up.

4 The half-life of americium-241 is 432 years. Why does this make it suitable as a source of alpha particles in a smoke alarm?

5 How could you use radioactivity to monitor the thickness of aluminium foil as it is being made?

6 Draw a flow chart to show how the thickness of paper is controlled.

7 Explain what properties a tracer used to detect leaks should have and what the drawbacks of using a tracer might be.

Learning Outcomes

6.3 Describe uses of radioactivity, including:
  a household fire (smoke) alarms
  d tracing and gauging thicknesses

HSW 12 Describe the benefits, drawbacks and risks of using new scientific and technological developments

These questions are indicative of the type of questions us in the exam. Refer to page 6 for information on the grad

Current, voltage and resistance

1. Thermal (heat) energy is produced when there is a current in a circuit.

   (a) (i) Which of these devices is designed to use the heating effect of an electric current?

   A   B   C   D

(1)

   (ii) Name a device in which this heating effect is a problem. Explain why it is a problem. (2)

   (b) (i) Ali set up this circuit.

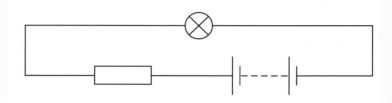

   Then he replaced the fixed resistor with a thermistor. Copy and change the circuit diagram to show this. (1)

   (ii) Which graph shows how the current varies with voltage for a thermistor?

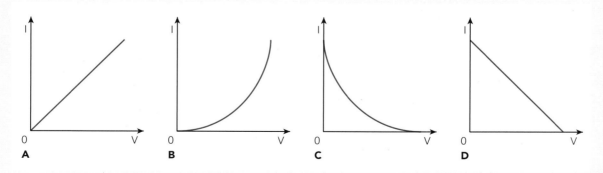

   A   B   C   D

(1)

   (iii) The temperature of the thermistor rises. Explain what happens to its resistance and to the current in the circuit. (2)

   (c) Explain, possibly by sketching a graph, how current varies with voltage for a diode. (3)

Swing-boat

2. Some friends played on a swing-boat ride like this.

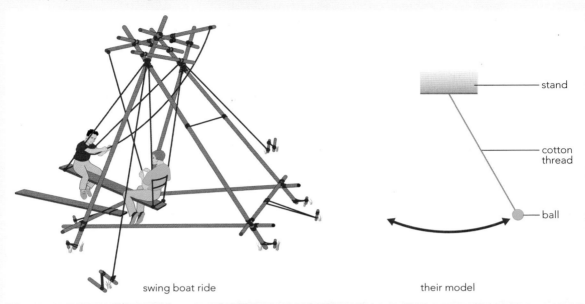

swing boat ride

stand

cotton thread

ball

their model

For a science project, Sophy tests the science using a model. She ties a piece of cotton thread to a ball and suspends it from a stand. Then she moves the ball from the vertical and lets it swing from side to side.

(a) Sophy measures the time it takes the ball to swing from one side to the other.
She does this for different lengths of cotton. Here are her results.

length of cotton (cm)	time of swing (s)
20	0.45
40	0.63
60	0.78

Why will the friends' swing-boat have much longer ropes than this? (1) ▬

(b) Complete these sentences by putting the correct word from the box into each of the spaces.
You may use each word once, more than once or not at all.

| kinetic   least   most   potential   zero |

The ball is pulled to one side and held stationary. It has _____ energy.

When it reaches the other side of the swing it has regained _____ energy.

At the centre of the swing it has most _____ energy.

The ball will finally stop swinging. Now the energy will be _____ energy. (4) ▬

(c) In Sophy's experiment, 4 J of work was done to move the ball from the vertical to one side. How much energy was transferred to the ball? (1)

(d) The 4 J of work was done in 0.5 s. How much power was used? Give the unit. (3)

(e) Suggest **two** factors that will affect the number of swings before the friends' swing-boat stops. (2)

Smoke alarms

3. Yosef ran a computer model to find out about americium-241 in his smoke alarm. He drew this graph to show how the amount of americium-241 changes with time.

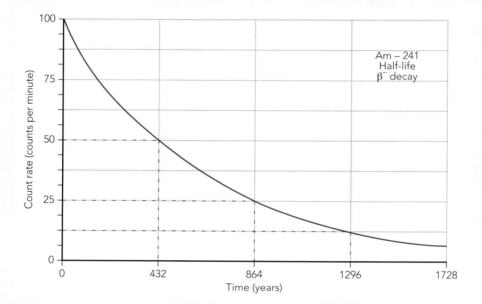

(a) The half-life of americium-241 is

A 216 years
B 432 years
C 500 years
D 864 years (1)

(b) Amira tells him that the smoke alarm will stop working when the activity of the americium-241 falls below 25 counts per minute. If Amira is correct, how long will the smoke alarm work for?

A 432 minutes
B 864 minutes
C 432 years
D 864 years (1)

(c) Give **two** reasons why it is better to investigate the americium using a computer model than to do the experiment yourself. (2)

(d) Americium-241 gives out alpha particles, which are the least penetrating type of ionising radiation. Smoke alarms are often fixed to the ceiling. How does this help to make them safe in normal use? (2)

(e) Old smoke alarms can be put into the normal rubbish for landfill sites. No more than 10 should be put in a bin bag. Suggest why this is a safe number. (2)

Electrostatic charges

4. A comb is rubbed on an insulating cloth. Both the comb and the cloth become charged. The comb has a negative charge.

(a) (i) What has happened to give the comb a negative charge? (2)

(ii) What type of charge is on the cloth? (1)

(iii) Compare the amount of charge on the cloth and on the comb. (1)

(b) The diagram shows two balloons hanging from the ceiling. They are hanging by cotton threads.

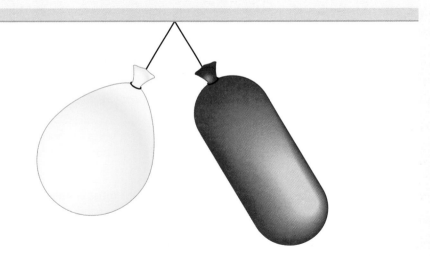

John suggests that the spherical balloon is charged and that the charge is positive. What evidence is there in the diagram to support John's two suggestions? (2)

(c) The photograph shows an aircraft being refuelled.

The aircraft is refuelled using a plastic fuel pipe. Explain how the plastic pipe can cause static electricity, which might result in an explosion. Also explain how a metal safety cable reduces the explosion risk.

(6)

Particles, ions and energy

5. (a) (i) Which of these has the largest mass?

   A a proton
   B a neutron
   C an alpha particle
   D a beta particle

   (1)

   (ii) Draw a line from each type of radiation to its correct properties.

mass number	4
charge number	−1

mass number	4
charge number	+2

alpha particle

mass number	0
charge number	−1

beta particle

mass number	0
charge number	+2

(2)

(iii) Which row of the table correctly compares alpha and beta particles?

	more ionising	more penetrating
A	alpha	alpha
B	alpha	beta
C	beta	alpha
D	beta	beta

(2)

(iv) Explain how a positive ion can be produced when an alpha particle passes through air. (2)

(b) Energy is released in a nuclear reactor when a uranium nucleus splits.

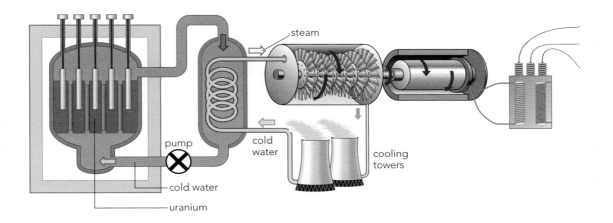

Describe the energy transfers and processes involved in converting the released energy into electrical energy. (6)

Lamps

1. Three lamps have different powers.
   They are connected in a circuit with a battery and three ammeters.

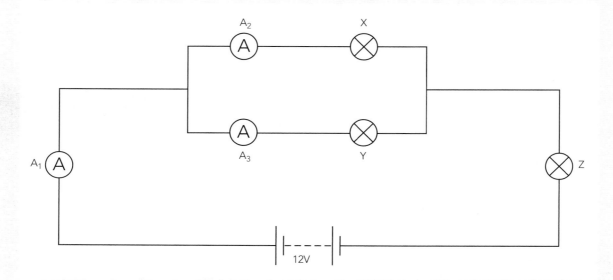

(a) (i) Lamps X and Y are described as being

     A   direct
     B   alternating
     C   in series
     D   in parallel
(1)

   (ii) Copy the diagram and show the connection of a meter to measure the potential difference across lamp Y.
(1)

  (iii) Ammeter A_1 shows a reading of 5 A and ammeter A_2 shows 2 A.
      Copy and complete the table to show the currents in lamps X, Y and Z.

lamp	Current (A)
X	
Y	
Z	

(2)

  (iv) Explain why a lamp glows when a current flows through it.
(2)

(b) A different lamp has a resistance of 960 Ω when the potential difference across it is 240 V.
Calculate the amount of energy transferred in 10 s as the lamp operates.
(4)

Charges in action

2. Some effects of electric charges have been known for thousands of years.

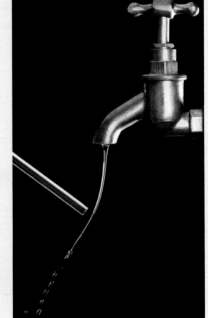

(a) (i) The following statements were made by famous scientists in the past.
Three are observations — one is a possible explanation.
Which is the explanation?

    A Rubbing a material for long enough can produce a spark
    B Electricity can come in two varieties that cancel each other
    C Some materials can become charged when rubbed, others can not
    D Electric attraction and repulsion can act across a vacuum   **(1)**

  (ii) Zachary bent a stream of water from a tap using a positively charged rod. He suggested that the water was attracted to the rod because it was charged as it emerged from the tap. Explain how he could test to see if the water was charged as it emerged from the tap.   **(2)**

(b) The air is removed from the space between two charged plates.
The plates are horizontal and one above the other. Some oil drops are sprayed into the space. If the drops are uncharged, they fall due to gravity.
One of the charged oil drops becomes stationary as shown.

  (i) What type of charge is on the oil drop?   **(1)**

positive plate

charged oil drop

negative plate

  (ii) There are two forces acting on the drop — the force of gravity and the force between charges. Draw a labelled free-body force diagram for when the drop is stationary.   **(2)**

  (iii) Explain what will happen to the drop if the amount of charge on each plate is increased.   **(2)**

  (iv) Explain what would happen if the charges on the plates were reversed.   **(2)**

Energy from the nucleus

3. One of the materials used in nuclear power stations is plutonium-239.
   A nucleus of this isotope can be represented by $^{239}_{94}Pu$.

   (a) (i) How many protons does an atom of this isotope contain?

           A 94
           B 145
           C 239
           D 333   (1)

       (ii) How many neutrons does an atom of this isotope contain?   (1)

       (iii) One isotope of plutonium is $^{239}_{94}Pu$. Suggest the atomic number and the mass number for a different isotope.   (2)

   (b) Explain the difference between fission and fusion.   (3)

   (c) Explain the purpose of the moderator in a nuclear reactor.   (2)

Falling

4. (a) Lynsey drops a small doll which has a toy parachute attached. She wants to find its acceleration. Name two items of basic equipment she would need.   (2)

   (b) In the first two seconds, the doll reaches a speed of 15 m/s.
   Calculate the average acceleration during this time. Give the unit.   (2)

   (c) This is an average acceleration because

           A it lasts for 2 s
           B it starts to fall from rest
           C it is changing continuously
           D it is less than the acceleration caused by gravity   (1)

   (d) The doll eventually reaches terminal velocity. Compare the forces on it now to when it started to fall.   (2)

   (e) Table tennis balls have a small mass. Imagine that identical table tennis balls are dropped from several metres high on Earth and on the Moon. They fall onto horizontal pieces of metal. Explain how and why the motion of the two balls, until they stop, will be different in the two places.   (6)

Radioactivity and nuclear waste

5. (a) Thorium-231 has a half-life of 25 hours. A sample contains 2 mg of thorium-231.
How much of this thorium will remain after 50 hours? **(2)**

(b) A solution containing a radioactive isotope is injected into a patient's body.
The radiation is monitored from outside the body. The chosen isotope should

	emit	have a half-life of
A	alpha	a few days
B	beta or gamma	a few days
C	alpha	a few years
D	beta or gamma	a few years

(1)

(c) A radioactive material X decays into another material Y. The graph shows how the amount
of Y produced from one sample of X changes with time if Y is stable.

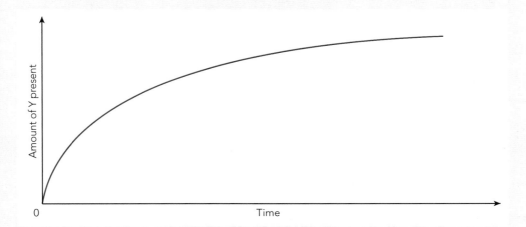

Imagine now that Y has a shorter half-life than X. Copy the graph and on the same axes,
sketch a line to show how the amount of Y will change with time. **(2)**

(d) The table gives details about some of the elements found in nuclear waste.

isotope	decays into	with a half-life of	and emits
xenon-137	caesium-137	3.8 minutes	beta
caesium-137	barium-137	30 years	beta
barium-137	stable	–	–
uranium-235	thorium-231	700 million years	alpha

Use data from the table to discuss how the dangers of ionising radiations affect the
techniques used to store and dispose of nuclear waste. **(6)**

Here are three student answers to the following question. Read the answers together with the comments around and after them.

| Question | Nuclear power stations | Grade | G–C |

The UK government plans to build ten new nuclear power stations to help meet the UK's future energy needs. Discuss the benefits **and** drawbacks of nuclear power stations. (6)

Student answer 1 — Extract typical of a level ① answer

Instead of 'clean' it would be better to say 'do not emit carbon dioxide'.

Uranium is correctly identified as a nuclear fuel, but the dangers of transporting it by road are not explained.

Nuclear power stations are clean but I don't like the look of them either. They give us reliable electricity. But they are very dangerous as radioactivity can kill and it is hard to store afterwards. Driving uranium on our roads is dangerous. Chernobill might happen again anywhere, anytime and kill lots of people.

Two drawbacks listed here, but some detail would be useful, e.g. radioactivity can cause cancer.

The nuclear disaster at Chernobyl is mentioned but no details are given.

Summary
The answer makes several points relating to the drawbacks of nuclear power. However, the points are very brief, lack detail and do not include scientific words or ideas. One benefit is mentioned ('Nuclear power stations are clean') but the reference is too unclear to get a mark.

Student answer 2 — Extract typical of a level ② answer

It would be good to say here that nuclear power stations do not cause climate change.

The answer says there was an explosion at Chernobyl but lacks detail.

Nuclear power stations have many benefits which make them seem to be the saviour of the world. They don't give out CO_2 or SO_2. They don't cause acid rain. They can produce continuous electricity 24-7 and do not rely on Sun or wind like other renewable fuels. On the other hand, many people are very worried by them since they think they give off radiation all the time. And the problem with explosions like at Chernobyl don't help. This was a serious problem which is still affecting the world today. Another major problem is the storing of nuclear waste. Sending it off into space or spreading it out in the ocean do not seem sensible. And, they don't really save energy since they take a lot of energy to be built in the remote parts of the country and to get the fuel.

Nuclear power is not actually renewable energy. Uranium supplies could eventually run out.

The answer mentions the problems of storing nuclear waste but does not give a scientific explanation.

Summary
The answer gives some good arguments on both sides of the debate. Sometimes the statements would benefit from more detail. There is very little detailed scientific knowledge in the answer.

A clear explanation of one of the main advantages of nuclear energy.

Nuclear power stations have several advantages. The biggest is the fact that they don't produce carbon dioxide. This is good because it means they don't add to global warming. Another benefit is they produce power all the time. Other kinds of clean energy, like wind power and solar, can't do this. Nuclear energy is very efficient too. It produces a lot of energy from a small amount of fuel. Nuclear fuel won't run out for many years. The biggest drawback is if there is an accident, nuclear power stations can blow up and spread radioactivity. This happened at Chernobyl in Russia. In Wales, farmers still cannot sell their sheep because of radioactivity from Chernobyl. The other big problem is the waste left once the fuel is used up. This waste is full of radioactive materials. Some of them have long half-lives. They are dangerous to humans for many years. So the waste has to be stored safely, and this is a big problem.

This answer is well organised: it sets out the advantages, then the drawbacks.

Evidence to back up the argument.

Good use of a scientific term.

Summary

This answer provides a balance of arguments on both sides of the question. Some scientific terms and explanations are also given. One possible area for improvement would be more detail on the risks of radiation to human health, for example increased risk of cancer. The answer also lacks a conclusion.

ResultsPlus

Move from level ① to level ②

To make progress from level 1 to level 2, your answer should give a better balance of arguments for and against nuclear power. A discussion question really requires arguments from both sides before reaching a conclusion. You should also include a little more science about, for example, the dangers of radioactivity and the problems with the storage of spent nuclear fuel.

Move from level ② to level ③

To move from level 2 to level 3, you should include a wider range of benefits and drawbacks of nuclear energy and include evidence to back up your arguments. Your discussion should try to present a balanced view before presenting a personal conclusion. Make sure that you clearly link such points as lack of carbon emissions and reduced global warming.

Here are three student answers to the following question. Read the answers together with the comments around and after them.

Question — Resistance

Grade: G–C

The voltage-current graphs for a piece of wire and a filament lamp are shown below.

The temperature of the wire is kept constant at 20°C. The temperature of the lamp filament is about 2500°C at 8 V.

Using the information above, explain the differences in resistance between the wire and the filament lamp.

You can work out resistance using the formula

$$R = \frac{V}{I}$$

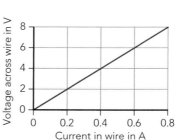

voltage-current graph for a wire

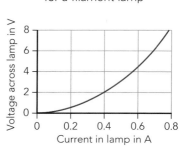

voltage-current graph for a filament lamp

Student answer 1 — Extract typical of a level ① answer

> The resistances of the graphs are different because resistance of a wire is fixed and the filament is shaped. The filament glows when it is hot. Electrical energy is changed into heat. More heat less current and voltage increases. The curve increases when voltage gets bigger.
>
> V = IR so 6 = 0.6 × R = 3.6R

This does not make it clear how resistance is related to the shape of each graph.

This calculation refers to the graph for the wire, but this is not explained in the answer. The calculation and units are also incorrect.

This and the previous sentence should be linked as both refer to the filament lamp.

'Increases' is not clear: it would be better to say that the curve becomes steeper.

Summary

This answer attempts to connect the resistance of the filament in the lamp to its temperature, but the connection is not made clearly. In discussing the graphs it fails to make clear which graph is being referred to. The student has used the formula to work out resistance incorrectly.

Student answer 2 — Extract typical of a level ② answer

> The differences in resistance between the wire and filament lamp is that one graph is straight and the other bent. The resistance depends on the shape of the graph. A straight line means a fixed resistance like a wire while a curve means the resistance changes like the filament.

'Curved' is a better word to use than 'bent'.

An important point: the resistance of the filament changes. It would also have been good to discuss the temperature change of the filament.

It is the slope of the *graph* that changes, not the filament.

> For the wire, the resistance is 8/0.8 which is 10 ohms. The slope of the filament increases, which means its resistance increases more and more as the current gets bigger and the wire heats up to 2500°C.

This links change in resistance and temperature. However, it is not made clearly and comes very late in the answer.

Summary

The answer is vague at the beginning, but improves later. The calculation of the resistance of the wire is correct but other calculations could have been made at other voltages/currents/temperatures. Some idea of why the filament becomes hot would be useful.

Student answer 3   Extract typical of a level ③ answer

This answer calculates the resistance at two points on the curve for the filament lamp. This gives clear evidence that the reistance is changing with temperature.

> The current/voltage graph for the wire is a straight line. It has a constant gradient, so the resistance is constant. This is because it is kept at constant temperature. Since resistance R = voltage V / current I the resistance, R, of the wire at 20 °C is 8/0.8 = 10 Ω for all values of current.
>
> The graph for the filament lamp shows that the resistance of the filament is constantly changing. At a current of 0.8 A, its resistance is also approximately 10 Ω (8/0.8 again). But, at 0.4 A the resistance is only 2/0.4 = 5 Ω. The reason for this is that the resistance changes with temperature. As the temperature rises (such as when there is a bigger current) there is friction with the flow of electrons. They are slowed down and their kinetic energy has gone into heat.

It is a good idea to define the symbols being used in an equation.

The answer gives an explanation of why temperature affects the filament's resistance.

Summary

In this answer the evidence from the graph is used to support a clear explanation of the fact that the resistance of the filament changes as it heats up. The descriptions of the graphs and what they show are mathematically correct.

 ResultsPlus

Move from level ① to level ②

To move to from level 1 to level 2 you should make it clear which graph you are describing at any time. Also you need to make the connection between the change in resistance of the filament and the change in its temperature. Practise calculations using electrical equations.

Move from level ② to level ③

To move from level 2 to level 3 you should make calculations using the formula. Instead of just mentioning the temperature differences, you should also show some understanding of the science behind why the filament's resistance changes as it heats up.

Here are three student answers to the following question. Read the answers together with the comments around and after them.

| **Question** | **Fusion and fission power stations** | **Grade** | **D–A*** |

The processes of nuclear fission and nuclear fusion both produce energy. There are 19 nuclear power stations in the UK. All of these use fission. However, the Government is investing a lot of money, time and effort to develop fusion power stations instead.

Explain why fusion power stations would be preferable but the UK only has fission power stations at present.

Student answer 1 — Extract typical of a level ① answer

Just repeating the question does not help to score marks.

Give specific names where you can, rather than general terms like 'materials', 'things' and 'that Russian one'.

This is good information but it does not explain why the sheep cannot be sold.

Fusion power stations would be preferable but the UK only has fission power stations at present because they are all we can make. Fusion is joining and fission is splitting. Fusion needs very high temperatures so the materials melt and can then be mixed so they join. The things produced in a fusion reaction are safe and the fuel for these reactors can be found in large amounts in the sea. Fission reactors produce materials which are very dangerous and need to be stored for a long time. Fission may go wrong and power stations explode sometimes like that Russian one. That is still affecting farmers in North Wales who cannot sell their sheep.

It is very important to say what is joining or splitting.

This could include more detail, e.g. explaining why the materials produced are dangerous and why they must be stored a long time.

Summary
This answer includes several important points but doesn't use the correct scientific terms or explain the points. Some topics that could be covered to gain extra marks include half-life linked to storage and mutation of cells caused by radiation.

Student answer 2 — Extract typical of a level ② answer

This first section does not provide any information that helps to answer the question.

Nuclear fission is a clean source of energy since it does not produce carbon dioxide. Using fusion would be similar. Fusion would be much better though because it does not produce radioactive materials. The materials used for joining in fusion are very small like hydrogen nuclei and waste products are also harmless. So they do not need to be stored and there is plenty of fuel in the sea. But there is a problem! The hydrogen nuclei both have the same charge and so repel each other.

The closer they get, the stronger they repel. For fusion, the nuclei must come into contact so they must be moving towards each other very, very fast. From the kinetic theory this means that fusion can only happen at very high temperatures a there has to be a high pressure to push them together. Scientists can do this, but cannot get more energy out than they put in. So fusion is no good for power stations yet.

Summary

This answer gives a very good description of the advantages and disadvantages of fusion. It gives plenty of detail and includes the correct scientific words. However, it includes hardly any information about fission.

 Extract typical of a level 3 answer

In fission the nuclei of a fuel such as uranium-235 is split into two smaller parts to release energy. This can be done at room temperatures and fairly safely. But the waste is very radioactive. Some will be dangerous for millions of years. So we have to store it so no radiation gets out. It even has to be safe from terrorists and earthquakes.
In fusion, the nuclei of small, light atoms such as hydrogen join together. The nuclei repel because they are both positively charged. Getting them to join together needs very high temperatures and pressures. Scientists can do this, but only for a very short time.
If we could get it to work, fusion would be much better than fission. We can get hydrogen from water. Also, the waste from fusion is harmless.

Summary

This answer is quite well-organised and gives all the most important information clearly. The final paragraph gives a clear summary of why we use fission, even though fusion would be better.

 ResultsPlus

Move from level 1 to level 2

Give more specific information, such as the names of nuclear fuels and other scientific terms. Copying out parts of the question in the answer will not get any marks. Give explanations for statements such as 'Fusion needs very high temperatures.'

Move from level 2 to level 3

Be more organised in preparing your answer. List the main points you want to make in rough and then order them. To compare two things you must describe them *both* and show similarities and differences. For this question the differences are more important than the similarities.

Here are three student answers to the following question. Read the answers together with the comments around and after them.

Question — Thinking and braking distances — Grade D–A*

On some motorways there are markings on the road. The markings are 40 m apart. Drivers are advised to keep at least two markings between cars.

Thinking and braking distances for different speeds are given in Figure 1. The maximum speed for a driver on the motorway is 70 mph. The thinking distance and braking distance can be affected by other factors, as well as speed.

Evaluate whether a distance of 40 m between markings is suitable.

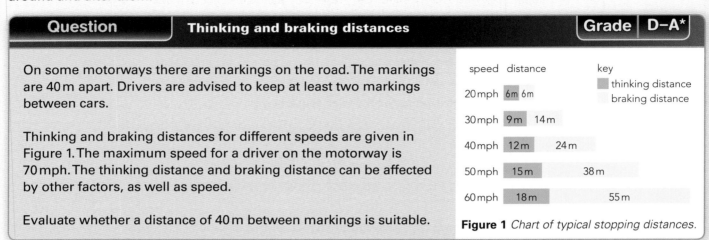

speed distance key
▇ thinking distance
☐ braking distance

20 mph 6 m 6 m
30 mph 9 m 14 m
40 mph 12 m 24 m
50 mph 15 m 38 m
60 mph 18 m 55 m

Figure 1 *Chart of typical stopping distances.*

Student answer 1 — Extract typical of a level ① answer

The minimum distance advised between cars is two markings. This is 80 m, not 40 m.

This is not related to the stopping distances given in Figure 1.

The 40 m is to allow for the driver drinking or if there is rain or snow. Old brakes or tyres will mean the car goes farther with the brakes on. Using a mobile phone is bad as well since UR not concentrating and u will go a long way before u react. 40 m is nearly as long as a football field. So it should be enough.

This suggests that drinking and driving is allowed. Both are extremely dangerous.

Text language and slang should not be used in answers.

Summary
The answer gives some ideas of the problems related to the driver and to the condition of the car. Problems with the weather and the road surface could have been noted. Relating the stopping distance to a football field gives a real-life reference but it would be better to refer to the thinking and braking distances shown in the chart provided as part of the question.

Student answer 2 — Extract typical of a level ② answer

Don't worry about the odd spelling mistake. Take care but marks will not be lost if there are one two errors.

40 m has been chosen to try to stop accidants. It takes a long time to stop. The stopping distance increases as the speed increases. At 60 mph the total stopping distance is 73 m. At 70 mph it would be more, so 80 m is about right. But if the driver is tired the thinking distance will be bigger than this meaning the 80 m will be too short.

Good point to make, though the student could also have said that the chart provides clear evidence for this.

The student could have included more factors that affect the thinking distance.

The same if the road is icy so the brakes don't work as well and it will take longer to stop. So 80 m might be OK but could be too short sometimes and the car will crash into.

The student could have included more factors that affect the braking distance.

Summary

What the student has written is good but more factors that affect thinking and braking distance could have been included. It would also have been good to make a more careful estimate of the thinking and braking distances at 70 mph. One point that is not mentioned is that the driver in front will also need time to think and then brake.

Student answer 3 | Extract typical of a level ③ answer

The markings on the road are so that you can brake to stop the car without hitting the car in front. The distance to stop a car depends on many things. Speed is the most important, but also things like how heavy the car is. The faster you are going the bigger the stopping distance. The stopping distance includes thinking distance as well as braking distance. The thinking distance is different for different people and can change for instance if you are tired. For 70 mph I think the total stopping distance is about 95 m. I got this by extending the pattern from the chart. This is more than 80 m but I think the car in front cannot stop immediately. So it should be OK in good conditions. The problem comes if it is raining or icy or oily, or if the car has old brakes or tyres. Then the braking distance will get bigger. The 80 m distance might not be enough.

Good explanation of the factors contributing to the stopping distance.

Making an estimate from the data supplied is good use of the data.

This answer lists a good range of factors that affect braking distance.

This is a point that has not been picked up in the other answers.

Summary

Evaluating something means presenting arguments for and against it, backed up by evidence. This answer uses the data supplied to show that the 80 m distance is enough in good conditions. On the other side of the argument it shows that in some cases the 80 m distance may not be enough.

ResultsPlus

Move from level ① to level ②

Include a greater range of factors which increase the stopping distance. Also, make use of the information supplied, in this case the chart of thinking and braking distances.

Move from level ② to level ③

Make sure that you include arguments for and against what is being evaluated. Give plenty of detail about the factors that affect thinking and braking distances. Also, estimate an actual value for the stopping distance at 70 mph and recognise that the car in front cannot stop instantly.

ResultsPlus

Phillip noticed that he received an electric shock from his car when he got out and closed the door. Explain why Phillip received a shock when he touched the door handle. (2)

Correct answer: Any two from this list would have got two marks:
- he had a static charge
- caused by friction
- that earthed when he touched the door handle.

(Or you could have explained that the car had become charged, and discharged through Phillip when he touched the handle.)

Most students got one mark for this, but not many got both. Common mistakes were:
- describing positive and negative electrons
- thinking that *both* Phillip and the car had a charge
- talking about attraction, repulsion or a reaction
- saying the charge came from friction between Phillip and the door handle.

ResultsPlus

The diagram shows a lorry carrying a crate. It is moving at a steady speed of 15 m/s. The mass of the crate is 200 kg. When the lorry stops suddenly, the crate continues to move forwards. Explain why. (2)

Correct answer: The crate has a lot of momentum (1) and there is not enough friction between the crate and the truck to stop it moving forwards (1).

Many students said that the crate was being pushed forwards and most did not mention friction and other resistive forces.

crate

direction of travel

ResultsPlus

In the nuclear reactor of a power station, neutrons hit uranium-235 nuclei. Describe what happens when a neutron hits a uranium-235 nucleus. (3)

Correct answer: Students needed to give three of the following points to get the three marks:
- a nucleus splits up (or breaks up)
- two daughter nuclei (two new elements) are formed
- two or more neutrons are also released
- energy is released.

About a third of students got all three marks and almost a quarter got two marks.

Nearly a fifth of students got no marks at all for this question. Answers that didn't get marks included:
- just talking about an atom 'breaking' without referring to splitting into two
- referring to 'daughter uranium' (the daughter nuclei are not uranium)
- referring to the daughter nuclei as 'particles' (they *are* particles, but this description isn't accurate enough to gain a mark), 'daughter atoms' or 'cells'.

You need to learn the words connected with fission and their meanings and use them accurately.

The diagram shows an osprey as it flies at a constant speed and at a constant height. Complete the diagram to show all the forces acting on the osprey.

▲ The correct answer is shown by the red arrows on the diagram. The labels point out important features of the correct answer.

■ Common mistakes in answering this question were:
- roughly drawn arrows (without using a ruler) that weren't horizontal or vertical
- arrows that weren't the correct length or lines without arrowheads or labels

lift

thrust

drag (or air resistance)

weight

The graph shows the velocity and time for a drag race.

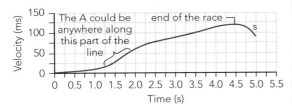

The A could be anywhere along this part of the line

end of the race

S

(a) Mark on the graph:
   (i) with an A, where the acceleration is largest (1)
   (ii) with an S, where the car is slowing down. (1)

▲ **Correct answer:** The text on the graph shows the correct answers.

■ Many candidates lost the mark for part (i) because their 'A' was not next to the steepest part of the line. Remember, the steeper the slope, the greater the acceleration.

(b) At the end of the race (as shown on the graph) the car has travelled 270 m. Calculate the average velocity of the car during the race. (2)

▲ **Correct answer:**   270 m/4.5 s (1)
                                            60 m/s (1)

■ Many put 270/5.0 = 54 m/s and lost a mark. They hadn't read the graph carefully enough. Although the line on the graph ends at 5.0 seconds, the race ends at 4.5 seconds and this is what the question was asking for. Some students even used 120 as the distance, even though the question said what the distance was. They had read the maximum value from the graph, forgetting that this is a velocity-time graph, *not* a distance-time graph.

Explain the term half-life. (1)
         Answer: The time it takes for half the atoms to decay (or the time for the activity to drop to half its original value).

How students answered

[   ] 0 marks

Many students lost the mark because they weren't careful enough with their wording. All these answers would *not* have gained a mark:
- The time it takes for an atom to decay.
- The amount of time an isotope has decayed.
- The time taken for the reactivity to halve.

[   ] 1 mark

Only half the students managed to get this mark.

acceleration A measure of how quickly the velocity of something is changing. It can be positive if the object is speeding up or negative if it is slowing down.

action In physics, one of a pair of forces. The reaction force acts in the opposite direction and on a different object.

active site The site on an enzyme molecule that has a special shape that holds the substrate molecule during the reaction.

active transport Movement of molecules into the cell using energy from respiration. It allows the cell to build up a high concentration of certain molecules inside the cell, against a concentration gradient. For example, the way that plant root cells take in mineral salts from the soil.

activity The number of emissions of ionising radiation from a sample in a given time. This is usually given in Becquerel.

actual yield The actual amount of product obtained from a chemical reaction. This is also known as the yield of the reaction.

adenine It is a base (a chemical found in DNA) and pairs up with thymine.

adult stem cell A stem cell found in differentiated tissue that can produce a few kinds of differentiated cell.

aerobic respiration Respiration that needs oxygen.

air bag Road safety device in which a bag suddenly inflates with gas to act as a cushion and reduce injury.

air resistance The force opposing the motion of an object moving through the air, sometimes called drag.

alimentary canal The muscular tube that runs from the mouth to the anus including the oesophagus, stomach, small intestine and large intestine.

alkali metal An element in group 1 of the periodic table.

allele Every gene comes in different types called alleles. So a gene for eye colour may come in a 'blue type' allele and a 'brown type' allele.

alpha particle The largest of the particles that can be emitted as radiation from an unstable nucleus. It consists of two protons and two electrons and is equivalent to a helium-4 nucleus.

alternating current Current whose direction changes many times each second.

ammeter An instrument for measuring the size of a current. It is put into a circuit in series with other components.

ampere (A) The unit for measuring current. 1 ampere is a flow of 1 coulomb of charge per second.

amylase A carbohydrase enzyme which breaks down starch to simple sugars.

amino acid A small molecule that is the building block of proteins.

anaerobic respiration Respiration that does not need oxygen.

anion Negatively charged ion.

antibodies Proteins that bind to the microorganisms that cause disease and destroy them.

aorta A major artery leading away from the heart.

aqueous solution Mixture formed when a substance is dissolved in water.

arteries Vessels that transport blood away from the heart.

asexual reproduction The formation of a new individual without fertilisation, using the process of mitosis to create offspring identical to the parent organism.

atom The smallest neutral part of an element that can take part in chemical reactions.

atomic number The number of protons in the nucleus of an atom. Also known as the proton number.

background count The average number of counts recorded by Geiger-Müller tube in a certain time from background radiation.

background radiation Ionising radiation that is around us all the time from a number of sources. Some background radiation is naturally occurring, but some comes from human activities.

balanced equation Description of a reaction using the symbols and formulae of the reactants and products, so that the number of 'units' of each element to the left of the arrow is the same as those to the right of the arrow.

barium meal A drink containing barium sulfate. It is swallowed so that the parts of the digestive system show up on X-ray photographs.

base The two strands in a molecule of DNA (in the double helix structure) are linked together at regular intervals by chemicals called bases. The bases are adenine, cytosine, thymine and guanine.

base pair (complementary base pair) The bases always pair up in the same way because of the complementary or matching shape of the molecules. Adenine (A) pairs up with thymine (T) and cytosine (C) pairs with guanine (G).

ⓗ base triplet A group of three bases that codes for a particular amino acid.

becquerel (Bq) The units for the activity of a radioactive object. One becquerel is one radioactive decay per second.

Benedict's test   Blue liquid which turns bright orangey-red when heated with simple reducing sugars.

beta-carotene   Substance in the human diet from which the body makes vitamin A.

beta particle   Particle of radiation emitted from the nucleus of a radioactive atom when it decays. It is an electron.

Bifidobacteria   Example of probiotic bacteria.

bile   An alkaline substance made by the gall bladder that helps in the digestion of fats.

bile duct   Tube that connects the gall bladder to the small intestine.

biodiversity   The variety of species present within a given area.

boiling point   Temperature at which a substance changes from a liquid to a gas.

bolus   A ball-shaped mass of chewed food.

bond   Force of attraction between atoms or ions.

blood vessels   Tube that contains the blood as it flows around the body.

braking distance   Distance travelled by a vehicle whilst the brakes are working to bring it to a halt.

bubble wrap   Plastic covering with many air bubbles made to protect delicate items.

by-products   Any products formed in a reaction in addition to the required product.

capillary   Tiny blood vessels with thin walls to allow diffusion of substances into and out of the blood.

carbohydrases   Enzymes which catalyse the breakdown of carbohydrates.

carbohydrate   Compound made up from carbon, hydrogen and oxygen, used for energy by organisms.

cardiac output   The volume of blood the heart can pump out in one minute, calculated using the equation cardiac output = stroke volume × heart rate

catalyst   A substance that speeds up the rate of a reaction without being used up in the reaction.

catalytic converters   Device fitted to car exhausts with a thin layer of transition metal catalyst on a honeycomb structure, giving a large surface area. The catalyst speeds up the reaction to combine carbon monoxide and unburned petrol in exhaust gases with oxygen from the air to form carbon dioxide and water vapour.

cation   Positively charged ion.

cell membrane   Thin layer which forms a semi-permeable barrier around the outer surface of the cytoplasm of the cell and controls the movement of substances into and out of the cell.

cells   The basic units of life in which many chemical reactions needed to sustain life (for example, growth) take place.

cellulose   Cell walls are made of tough cellulose which support the cell and allow it to keep its shape.

cell wall   Relatively rigid structure that surrounds plant and bacterial cells, which support the cell and help it to keep its shape.

chain reaction   The sequence of reactions produced when a nuclear fission reaction triggers one or more further fissions.

chlorophyll   Green substance in chloroplasts that absorbs energy from sunlight.

chloroplast   Organelle in plant cells that contains chlorophyll, and is where photosynthesis takes place.

cholesterol   A fat which is made in the liver and carried around the body in the blood. High levels are associated with an increased risk of heart disease.

chromatogram   The piece of paper showing the results of carrying out chromatography on a substance.

chromatography   A technique for separating the components of a mixture, for example different food coloured colouring agents. It works because dissolved substances move at different rates through a piece of paper soaked in a solvent.

chromosomal DNA   The DNA which makes up the chromosomes of a cell.

chromosome   Thread-like structures found in the nucleus of the cell which carry the genetic information.

circulatory system   Organ system involving the heart and blood vessels which oxygenates blood and moves it around the body.

clone   An individual created by a form of asexual reproduction to produce offspring that is genetically identical to the parent.

Ⓗ **codon**   Another name for a base triplet.

collision theory   The theory of chemical reactions that describes how particles must collide with enough energy to react.

complementary base pair   See base pair.

compound   A substance containing two or more elements chemically joined together.

compound ion   A group of atoms that together have a positive or negative charge.

concentration   A measure of how much solute is dissolved in a solvent. The more solute that is dissolved in a fixed volume of solvent, the higher the concentration of the solution.

concentration gradient A situation when a solute is in a gradually increasing concentration from one region to another.

conservation of energy The fact that energy can be transferred, but the total amount of all forms of energy before is equal to the total afterwards.

conserved A quantity is kept the same throughout e.g. momentum, energy.

control rod A rod that can be lowered into the core of a nuclear reactor, absorb neutrons and slow down the chain reaction.

core The part of a nuclear reactor where controlled fission takes place.

cosmic ray Charged particles with a high energy that come from stars, neutron stars, black holes and supernovae.

coulomb (C) The unit for measuring charge.

count rate The number of alpha or beta particles or gamma rays detected by a Geiger-Müller tube in a certain time.

covalent bond The bond formed when a pair of electrons is shared between two atoms.

crumple zone A vehicle safety device in which part of the vehicle is designed to crumple in a crash, reducing the force of impact.

current A movement of electrons (or other charged particles).

cytoplasm The liquid gel which makes up a lot of the body of a cell and is where many chemical reactions take place.

cytosine It is a base (a chemical found in DNA) and pairs up with guanine.

daughter cell Cell produced from division of a parent cell.

daughter nucleus A nucleus produced when the nucleus of an unstable atom splits into two during fission or when a radioactive nucleus decays by emitting an alpha or beta particle.

decays See radioactive decay.

delocalised electrons Free electrons that can move around between ions in a metal (H) or in the layers of graphite)

denature To break down/change shape, as proteins denature with excess heat.

deoxygenated Without oxygen.

deoxyribonucleic acid (DNA) that makes up genes and chromosomes; the instructions for a cell's growth and activity.

diamond A form of carbon that is very hard and doesn't conduct electricity. It is used in cutting tools.

differentiate Specialise, develop into different kinds, as in cells that become nerve, muscle or bone cells.

differentiation The process by which a less specialised cell becomes more specialised for a particular function. The cell normally changes shape to achieve this.

diffusion The random movement and spreading of particles. When there is a concentration gradient, there is a net (overall) diffusion of particles from areas of high concentration to regions of lower concentration.

digestion The breakdown of large food insoluble molecules into small, soluble food molecules.

digestive system The system of organs which brings about digestion of the food in the body.

diode A component that lets electric current flow through it in only one direction.

diploid A cell that has two sets of chromosomes. In humans, almost all cells except the sperm and egg cells are diploid.

direct current Electric current that flows in the same direction all the time.

discharge To remove an electric charge by conduction.

displacement The distance travelled in a particular direction. Displacement is a vector, distance is not.

displacement reaction When a more reactive substance displaces a less reactive substance from one of its compounds.

dissolve This occurs when a solute splits up and mixes with a solvent to make a solution.

distance How far something has travelled.

distance–time graph A graph of the distance travelled against time for a moving object. The slope on a distance–time graph gives the speed of a body.

distribution The places in which a certain organism can be found in an area.

DNA replication When the chromosomes are copied before cell division occurs.

dot and cross diagram A way of showing electronic structures in covalent substances, using dots and crosses to represent the electrons from different atoms.

double bond When two bonds are joined by two bonds.

double helix The spiral structure of a DNA molecule, produced by two strands of joined by complementary base pairs.

drag Another name for air resistance.

earth [electricity] To connect something to earth, so that any electrostatic charge can flow away.

ecosystem An area in which all the living organisms and all the non-living physical features in the area form a stable relationship that needs no input from outside the area to remain stable.

electrolysis   The process in which electrical energy, from a d.c. supply, decomposes some compounds.

electron   Negatively charged subatomic particle found in shells around the nucleus of an atom. An electric current in a metal wire consists of moving electrons.

electron microscope   Instrument which magnifies specimens using a beam of electrons.

electronic configuration   The arrangement of electrons in shells around the nucleus of an atom.

electrostatic charge   An overall electric charge caused by an object gaining or losing electrons.

electrostatic repulsion   A force between two electrical charges that have the same sign that pushes them apart.

element   Substance that cannot be split up into simpler substances.

elongation   Getting longer.

embryo   The ball of cells produced by cell division of the zygote. A very early stage in the development of a new individual.

embryonic stem cell   A cell from an early stage of division of an embryo that can produce almost any kind of differentiated cell.

empirical formula   The simplest whole number ratio of atoms of each element in a compound.

Ⓗ emulsify   Turn into an emulsion, a mixture in which particles of one liquid are suspended in another liquid.

endothermic   Reaction that takes heat energy in, decreasing the temperature of the reaction mixture and its surroundings.

energy level   See shell.

energy transfer   Energy being moved from one place to another, possibly with a change in energy form at the same time.

Ⓗ enucleate   Remove the nucleus from a cell.

environment   An organism's surroundings; made up of many different factors such as air, water, soil and other living organisms.

enzyme   A protein molecule made by living cells that speeds up the rate of a reaction.

evolution   The development of new species over time through a process of natural selection.

excess post-exercise oxygen consumption (EPOC)   Period of time after exercise in which a greater than usual amount of oxygen is needed by the body for various processes, including the removal of lactic acid cell repair. This used to be known as the oxygen debt.

exothermic   Reaction that releases heat energy, increasing the temperature of the reaction mixture and its surroundings.

explosions   Any situation in which two or more objects, initially stationary and in contact, move apart from one another.

faeces   Undigested, waste material.

fat   Chemicals that are used to store energy in organisms.

Ⓗ fatty acid   Part of the structure of a fat or oil.

fertilise   When two gametes fuse.

filament lamp   A light bulb whose resistance increases when it gets hot.

filtration   The separation of undissolved solids from a liquid by filtering.

flagella   Whip-like protein bodies found on the outside of bacterial cells and other cells which beat back and forwards and can be used for movement.

flame test   An analytical test to find out which metal ion is present in a substance. Different metals produce different colours in a Bunsen burner flame.

force   An interaction on an object that can cause it to accelerate.

formula   Abbreviation for a substance using symbols with two or more atoms or ions.

fossil   The preserved traces or remains of an organism which lived a very long time ago.

fossil record   The collection of fossils identified from different periods of time that can be interpreted to form a hypothesis about the evolution of life on Earth.

fraction   A component of a mixture that has been separated by fractional distillation.

fractionating column   A long column used to separate a mixture of liquids into different fractions with different boiling points. The column is warmer at the bottom than at the top.

fractional distillation   The process by which a mixture of two or more liquids is separated due to their different boiling points.

free-body diagram   A diagram of an object showing all the forces acting upon it and the size and direction of those forces.

friction   A force between two surfaces that resists motion and is always opposite to the direction of a moving object.

functional foods   Foods which are not eaten for nutritional value but which claim to make you healthier.

gall bladder   An organ that stores the bile made by the liver and releases it into the small intestine via the bile duct.

gamete   Sex cell (sperm or egg cells), produced in the male or female reproductive organs by meiosis.

gamma ray   A high-frequency electromagnetic wave emitted from the nucleus of a radioactive atom. Gamma rays have the highest frequencies in the electromagnetic spectrum.

gas exchange   A process in the lungs in which oxygen diffuses from the lungs into the blood and carbon dioxide diffuses from the blood into the lungs.

Geiger-Müller tube   A device that can detect ionising radiation and is used to measure the activity of a radioactive source.

gene   A section of DNA which codes for a specific protein

genetic code   The code produced by the sequence of bases in genetic material (e.g. DNA).

genetic engineering   The process of removing a gene from one organism and inserting it into the DNA in a cell from another organism.

genetically modified organisms (GMOs)   An organism that has had a gene from another species introduced (e.g. inserting a gene for producing human insulin into a bacterium).

genome   All of the genetic information (DNA) of an organism, as a list in order of every base.

giant molecular, covalent substance   Substance containing millions of atoms all held together by covalent bonds.

golden rice   Genetically engineered rice which produces beta-carotene in the rice grains turning them a golden yellow colour.

glucose   A simple sugar that is broken down in cells to release energy during respiration. It is also produced during photosynthesis.

(H) glycerol   Part of the structure of a fat or oil.

gradient   A measurement of the steepness of the slope of a graph. The steeper the graph, the higher the gradient.

graphite   A form of carbon that is soft and conducts electricity. It is used as a lubricant as the layers slip over each other easily.

gravitational field strength   A measure of how strong the force of gravity is somewhere.

gravitational potential energy   The energy stored in things that can fall.

group   A column in the periodic table, containing elements with similar properties.

growth   Increase in size, length and mass, as well as increase in cell number.

guanine   It is a base (a chemical found in DNA) and pairs up with cytosine.

habitat   The place where an organism usually lives.

haemoglobin   The red iron-containing pigment found in red blood cells.

half-life   The average time taken for half of the radioactive nuclei in the original sample to have decayed.

halide   A compound formed between a halogen and another element, such as a metal or hydrogen.

halogen   An element in group 7 of the periodic table.

haploid   Having one set of chromosomes, as in gametes.

herbicide   Chemical which kills plants, usually used on weeds.

high-level waste (HLW)   Highly radioactive waste which produces large amounts of ionising radiation. The radioactivity decreases over a few tens of years and it becomes intermediate level waste.

Human Genome Project (HGP)   A project to sequence (order) all of the base pairs of the human genome involving scientists from many different countries working together.

hydrogen bond   Base pairs are joined together by weak hydrogen bonds.

immiscible   Liquids that do not form a solution but separate into two layers.

(H) implant   In reproduction, placing an embryo into the uterus of a female animal to develop.

in parallel   Components connected so that the current splits up. Some of the current will flow through one component, and some will flow through the other.

in series   A component connected in a circuit so that current must pass through it to the rest of the circuit.

induction   When an object is charged by another charged object placed close to it.

induced charge   An electrostatic charge caused by another charged object being brought near.

inert   Does not react.

insoluble   Substance which does not dissolve in a given solvent.

interact   When two objects interact, there is always a pair of forces. The forces always act on different objects – one force on each object involved in the interaction.

intermediate-level waste (ILW)   Materials which have become radioactive because they have been in a nuclear reactor. It remains radioactive for tens of thousands of years.

interpolation   Estimating the value of a new data point on a scatter graph between the values already known

ion   An atom or group of atoms with an electrical charge. Atoms become positively charged ions if they lose electrons and negatively charged ions if they gain electrons.

ionic bond   Strong electrostatic force between oppositely-charged ions.

ionic compound   Substance containing ions from two or more elements.

ionising radiation   Form of radiation that causes ionisation in atoms of the material it passes through. The atoms lose or gain electrons and become charged – they become ions.

irradiate   To use powerful doses of gamma rays to sterilise food or medical equipment.

isotopes   Atoms of an element with the same number of protons and electrons but with different numbers of neutrons.

joule (J)   Unit for energy, and work done.

kinetic energy   Movement energy.

lactic acid   The waste product of anaerobic respiration in animal cells.

Lactobacillus   Example of probiotic bacteria.

large intestine   Organ that absorbs water from digested material.

lattice structure   Regular, grid-like arrangement of particles, such as ions.

left atrium   One of the four chambers of the heart that receives blood from the pulmonary vein.

left ventricle   One of the four chambers of the heart that receives blood from the left atrium and pumps it into the aorta.

light dependent resistor (LDR)   A resistor whose resistance gets lower when light shines on it.

light microscope   Instrument which magnifies specimens using light and lenses.

limiting factor   A single factor that when in short supply can limit the rate of a process such as photosynthesis.

lipase   Enzyme which digests fats to fatty acids and glycerol.

liquefy   Convert a substance into a liquid by heating or cooling.

liver   Organ that has a range of functions including secretion of bile.

lock-and-key hypothesis   An idea that describes the relationship of a substrate and the active site of an enzyme to help explain how enzymes work.

low-level waste (LLW)   Slightly radioactive waste, usually clothing, cleaning materials and medical equipment.

lubricant   A substance placed between two moving surfaces to reduce the friction between them.

malleable   Can be hammered into shape.

mass   A measure of the amount of material that there is in an object.

mass number   The total number of protons and neutrons in the nucleus of an atom. Also known as the nucleon number.

meiosis   Division of parent cell that produces genetically different haploid cells.

melting point   Temperature at which a substance changes from a solid to a liquid.

🅗 messenger RNA (mRNA)   The molecule formed during DNA transcription that carries the code from the chromosome to a ribosome.

metallic bond   The type of bonding in metals.

miscible   Liquids that completely mix together to form a solution.

mitochondrion (plural mitochondria)   The site of cellular respiration where glucose is broken down using oxygen to release energy, which is needed for reactions in the cell.

mitosis   Division of a parent cell that produces two genetically identical diploid cells.

moderator   A substance in a nuclear reactor which slows down neutrons so that they can be absorbed by the nuclear fuel more easily.

molecular formula   This shows the actual number of atoms of each element that combine to make a molecule of a compound.

molecule   Two or more atoms joined together by covalent bonds.

molten   Melted to form a liquid.

momentum   measure of motion, mass multiplied by velocity.

mutation   A change in the base sequence of DNA (often as a result of exposure to radiation)..

neutron   Electrically neutral subatomic particle found in the nucleus of most atoms.

noble gases   Elements in group 0 of the periodic table.

nuclear fission   The reaction when the nucleus of a large atom, such as uranium, splits into two smaller nuclei.

nuclear fusion   The reaction when the nuclei of light atoms, such as hydrogen, combine to make the nucleus of a heavier atom.

nuclear reactor   The part of a nuclear power station that contains the fuel rods, control rods and moderator and coolant.

nucleon   The sub-atomic particles in the nucleus of an atom, i.e. protons and neutrons.

nucleon number   The number of nucleons (protons and neutrons) in a nucleus.

nucleus In chemistry, the positively charged centre of an atom. In physics, the central part of an atom, containing protons and neutrons. In biology, contains DNA for making new cells and organisms and also controls the reactions in the cell.

oesophagus Muscular tube between the mouth and stomach.

ohm (Ω) The unit for measuring resistance.

oligosaccharides A type of carbohydrate which is a common prebiotic

organ A group of different tissues working together to carry out a particular function.

organelles Tiny structures that carry out specific jobs, for example, nucleus and mitochondria.

organ system A group of organs working together to carry out a particular function in the body.

osmosis The diffusion of water from a region of high concentration of water molecules to a region of lower concentration of water molecules, through a partially permeable membrane.

oxygenated With oxygen.

oxyhaemoglobin The compound formed when haemoglobin combines with oxygen in the lungs. The form in which oxygen is carried around the body.

pancreas Organ that makes digestive enzymes and secretes them into the first part of the small intestine.

parallel See in parallel.

parallel circuit A circuit in which there is more than one path for the current to follow.

parent cell The cell that divides to produce daughter cells

partially permeable membrane A thin sheet of material that will allow certain small molecules to diffuse through it (e.g. water) but not other larger ones.

peer review When a scientist who has a similar scientific background and experience checks someone else's work.

penetration distance The distance ionising radiation can go through a substance.

pentadactyl Five fingered.

pepsin An example of a protease enzyme found in the stomach.

percentage by mass The percentage of total mass for a specific element in a compound; found from the relative mass of the element in the compound divided by the relative formula mass of the compound and expressed as a percentage.

percentage yield The actual yield divided by the theoretical yield as a percentage.

percentile The value of a variable below which a certain percentage of observations fall. For example the 20th percentile of an ordered set of data indicates that 20% of the data points are the same or lower than this value.

period row in the periodic table in which the atomic number increases by one from one element to the next.

periodic table Chart in which the chemical elements are arranged in order of increasing atomic number.

peristalsis The waves of muscular contraction that move food along the alimentary canal.

phloem Living tissue that transports sugars around a plant.

photosynthesis A series of enzyme-catalysed reactions carried out in the green parts of plants. Carbon dioxide and water combine to form glucose. This process requires light energy from sunlight.

pitfall trap A trap used to catch small animals that move on the ground. The animals are unable to escape.

plant stanol esters Oily substances found in plants that appear to lower blood cholesterol levels in people.

plasma The liquid component of the blood that carries all the suspended cells and dissolved substances.

plasmid/ plasmid DNA A circle of extra DNA found only in bacteria cells.

platelets Cell fragments that are important in the clotting mechanism of the blood.

Ⓗ polypeptide A chain of amino acids that will form part of a protein.

pond net A net used to collect aquatic organisms from ponds, rivers and streams.

pooter A simple device used to collect small invertebrates.

population size The numbers of individuals of a species in an area.

potential difference Another word for voltage. It is the difference in the energy carried by electrons before and after they have flowed through a component.

potometer A device used for measuring the rate of water uptake by a plant.

power The amount (rate) of energy transferred per second. The units are watts (W).

prebiotics Substances which cannot be digested by human digestive enzymes but which act of food for probiotic bacteria in the intestine.

precipitate Insoluble product formed in a precipitation reaction.

precipitation reaction   Reaction in which an insoluble product is formed from soluble reactants.

probiotics   Foods containing live bacteria that produce lactic acid in the gut and may improve the health of your digestive system.

protease   Enzyme which digests proteins to amino acids.

protein   A polymer made up of amino acids, containing carbon, hydrogen, oxygen and nitrogen. Genes carry the instructions for making proteins.

protein synthesis   The building up of a protein molecule by joining together amino acids.

proton   Positively charged subatomic particle found in the nucleus of all atoms.

proton number   The number of protons in the nucleus of an atom. Also known as the atomic number.

pulmonary artery   Arteries that carry deoxygenated blood from the right ventricle to the lungs.

pulmonary vein   Veins that carry oxygenated blood from the lungs to the left atrium.

quadrat   A square frame of known area, such as 1 m^2, which is placed on the ground to get a sample of the organisms living in a small area.

radioactive   A substance that gives out ionising radiation such as alpha or beta particles or gamma rays is radioactive.

radioactive decay   When an unstable nucleus changes by giving out ionising radiation to become more stable.

radioactive waste   Material left over after the fission of uranium that is radioactive.

radiotherapy   Use of ionising radiation to treat diseases, such as to kill cancer cells.

radon   Naturally occurring radioactive gas that is emitted from rocks under the ground as a result of decay of radioactive uranium. In some parts of the UK, it is emitted at dangerous levels.

random   Any process that cannot be predicted and can happen at any time is said to be random.

random sampling   A method of sampling where the locations are selected randomly.

rate of a chemical reaction   The speed at which a chemical reaction takes place, as shown by the change in the amount (or the concentration) of a reactant (or product) with time.

reaction   In physics, one of a pair of forces. This acts in the opposite direction and on a different object to the action force.

reaction time   The time taken to respond to a stimulus, which is affected by the speed of activity in the brain and nervous system.

reactivity   How fast or how readily an element reacts.

red blood cells   Biconcave discs containing haemoglobin that give the blood its red colour and carry oxygen around the body to the tissues.

relative abundance   The number of objects of a particular kind in a sample shown as a percentage of the total number of objects in the sample.

relative atomic mass   The mean mass of an atom relative to the mass of an atom of carbon-12, which is assigned a mass of 12.

relative charge   The electrical charge of a subatomic particle compared to the electrical charge of a proton.

relative formula mass   The sum of the relative atomic masses of all the atoms in a formula.

relative mass   The mass of a subatomic particle compared to the mass of a proton.

representative sample   A sample that has approximately the same characteristics as those of the whole study area.

resistance   How easy or difficult it is for an electric current to flow through something.

respiration   A series of reactions occurring in all living cells in which glucose is broken down to release energy.

resultant force   The total force that results from two or more forces acting upon a single object. It is found by adding together the forces, taking into account their directions.

R_f value   The ratio of the distance travelled by a solute on a chromatogram (measured from the centre of the spot) to the distance travelled by the solvent under the same conditions. Also called retardation factor. The R_f value for a substance can be used to identify compounds.

H ribosome   Small structures in the cytoplasm of a cell where mRNA is translated into an amino acid chain.

right atrium   One of the four chambers of the heart that receives blood from the vena cava.

right ventricle   One of the four chambers of the heart that receives blood from the right atrium and pumps it into the pulmonary artery.

root hair cells   Cells found near the tip of roots that have thin extensions (that look a bit like hairs). Water enters root hair cells by osmosis. The long thin extension gives a large surface area.

saliva   Lubricates food and makes it easier to swallow, also contains amylase which begins digestion of carbohydrate.

salt A compound formed by neutralisation of an acid by a base. The first part of the name comes from the metal in the metal oxide, hydroxide or carbonate. The second part of the name comes from the acid.

sampling Looking at a small portion of an area or population.

seat belt Vehicle safety device in which a material strip holds a person in place within a vehicle. It is designed to stretch in an impact, reducing the force of impact.

separating funnel A funnel with a tap at the bottom that is used for separating immiscible liquids.

septum A thin membrane that separates the two sides of the heart.

series See in series.

series circuit A circuit in which there is only one loop for the current to flow through, so the current goes through all the components one after another.

sexual reproduction The formation of a new individual from the fertilisation of a female gamete (egg cell) by a male gamete (sperm cell). This individual is genetically different from its parents.

shell In chemistry, space around a nucleus that can be occupied by electrons, usually drawn as a circle. Also referred to as an energy level.

shielding Prevents ionising radiation and stray neutrons escaping from the nuclear reactor.

simple molecular, covalent substance A substance made of individual small molecules, with strong covalent bonds holding the atoms together in the molecules but only weak forces between neighbouring molecules.

small intestine Organ where digestion is completed and nutrients are absorbed.

solubility A measure of the amount of substance that will dissolve in a certain volume of solvent.

soluble Substance which dissolves in a given solvent.

solute The solid or liquid that dissolves in a given solvent to form a solution.

solution The clear mixture that forms when a solute dissolves in a given solvent.

solvent The liquid that dissolves the solute. solute + solvent → solution

specific Only one, as in an enzyme only catalyses the reaction of one kind of substrate.

spectroscopy Analysis of the light on a spectrum (e.g. of the wavelength of visible light emitted by atoms in flame tests).

speed A measure of the distance an object travels in a given time. Usually measured in metres per second (m/s).

starch A carbohydrate, made by joining together thousands of glucose molecules.

state symbol Letter or letters to show the state of a substance (e.g. (g) for gas).

static electricity Electric charges on insulating materials.

stem cell An unspecialised cell that can divide to produce more stem cells or different kinds of specialised cell.

sterilise To destroy bacteria, viruses, mould and pests such as insects on an object. It can be carried out using radioactive sources.

stomach Organ that makes acid and some enzymes.

stomata (singular stoma) A tiny pore in the lower surface of a leaf, which when open, allows gases o diffuse into and out of the leaf.

stopping distance The sum of the thinking and braking distances.

stroke volume The volume of blood the heart can pump out with each beat.

sub-atomic particle A particle that is smaller than an atom, such as a proton, neutron or electron.

substrate The substance that is changed by an enzyme in a chemical reaction.

sugars A group of compounds formed from carbon, hydrogen and oxygen.

surface area The total area of all the surfaces of an object or substance.

surface area to volume ratio The total amount of surface area of an object divided by its volume. The surface area to volume ratio of a small object or organism is larger than that of a large object of a given shape, for example a cube. The surface area to volume ratio of a wide and thin structure, or a wrinkled structure, is greater than that of a spheroidal one.

(H) surrogate mother A female who is not related to the embryo that is implanted in her uterus to develop.

sweep net A net used to collect insects from long grass or the canopy of a tree, by "sweeping" it through the grass or leaves.

systematic sampling A method of sampling where the samples are selected from the population at regular or systematic intervals, (e.g. a sample every metre or every fifth person).

temperature A measure of how hot something is.

terminal velocity A constant, maximum velocity reached by objects falling. This happens when the weight downwards is equal to the air resistance upwards.

theoretical yield   The maximum calculated amount of a product that could be formed from a given amount of reactants.

thermistor   A component whose resistance gets smaller when it gets hotter.

thinking distance   Distance travelled by a vehicle whilst the driver reacts.

thymine   It is a base (a chemical found in DNA) and pairs up with adenine.

tissue   A group of specialised cells that all carry out the same function.

toxic   Poisonous.

tracer   A radioactive substance that is deliberately injected into the body or into moving water. It allows the movement of the substance to be followed by detecting the ionising radiation emitted.

(H) transcription   When a strand of mRNA is produced by complementary pairing of bases with one strand of DNA in the nucleus.

transfer   Energy changing form.

(H) transfer RNA (tRNA)   A small RNA molecule that transfers the correct amino acid to the ribosome during translation so that the protein it codes for can be synthesised.

transition metals   The metals in the central block of the periodic table.

(H) translation   Transferring the code in mRNA sequence into a sequence of amino acids on a ribosome.

transpiration   The evaporation of water vapour from the surface of a plant.

unstable   An unstable nucleus is one that will decay and give out ionising radiation.

(H) uracil   A base only found in RNA, which replaces the base thymine found in DNA.

(H) uterus   Womb.

vacuole   Membrane-bound space in the cytoplasm filled with cell sap, a store of water and nutrients. It helps to support the plant by keeping the cells rigid.

validate   To confirm that a scientific theory is true.

valves   Flaps of tissue in the heart that stop the blood flowing backwards.

variable resistor   A resistor whose resistance can be changed by turning a knob or moving a slider.

vector   A quantity that has a magnitude (size) and direction. Force and velocity are examples of a vector. Speed, mass and volume are not vectors.

vector quantity   Quantity that has both a size and a direction.

veins   Vessels that transport blood back to the heart.

velocity   The speed of an object in a particular direction. Usually measured in metres per second (m/s).

vena cava   A major vein leading to the heart.

villi   Fingerlike folds of the lining of the small intestine which greatly increases the surface area for the absorption/diffusion of digested food products into the blood.

visking tubing   Synthetic membrane which is partially permeable – small molecules can pass through but not larger molecules.

voltage   Another word for potential difference. A way of measuring the amount of energy transferred to a component by a current.

voltmeter   Instrument for measuring the potential difference across a component.

waste product   By-products that have no uses.

watt (W)   The unit for measuring power. 1 watt = 1 joule of energy transferred every second.

weight   The force pulling an object downwards, it depends upon the mass of the object and the gravitational field strength.

white blood cells   Several different types of cells that are all part of the body's defence system against disease.

work   Transfer of energy.

xylem   Tissue made of dead hollow cells that transports water and dissolved minerals from the roots to other parts of the plant.

yield   The amount of product formed in a reaction.

zygote   A fertilised egg cell.

1	2		3	4	5	6	7	0
								4 **He** helium 2
7 **Li** lithium 3	9 **Be** beryllium 4		11 **B** boron 5	12 **C** carbon 6	14 **N** nitrogen 7	16 **O** oxygen 8	19 **F** fluorine 9	20 **Ne** neon 10
23 **Na** sodium 11	24 **Mg** magnesium 12		27 **Al** aluminium 13	28 **Si** silicon 14	31 **P** phosphorus 15	32 **S** sulfur 16	35.5 **Cl** chlorine 17	40 **Ar** argon 18

39 **K** potassium 19	40 **Ca** calcium 20	45 **Sc** scandium 21	48 **Ti** titanium 22	51 **V** vanadium 23	52 **Cr** chromium 24	55 **Mn** manganese 25	56 **Fe** iron 26	59 **Co** cobalt 27	59 **Ni** nickel 28	63.5 **Cu** copper 29	65 **Zn** zinc 30	70 **Ga** gallium 31	73 **Ge** germanium 32	75 **As** arsenic 33	79 **Se** selenium 34	80 **Br** bromine 35	84 **Kr** krypton 36
85 **Rb** rubidium 37	88 **Sr** strontium 38	89 **Y** yttrium 39	91 **Zr** zirconium 40	93 **Nb** niobium 41	96 **Mo** molybdenum 42	[98] **Tc** technetium 43	101 **Ru** ruthenium 44	103 **Rh** rhodium 45	106 **Pd** palladium 46	108 **Ag** silver 47	112 **Cd** cadmium 48	115 **In** indium 49	119 **Sn** tin 50	122 **Sb** antimony 51	128 **Te** tellurium 52	127 **I** iodine 53	131 **Xe** xenon 54
133 **Cs** caesium 55	137 **Ba** barium 56	139 **La*** lanthanum 57	178 **Hf** hafnium 72	181 **Ta** tantalum 73	184 **W** tungsten 74	186 **Re** rhenium 75	190 **Os** osmium 76	192 **Ir** iridium 77	195 **Pt** platinum 78	197 **Au** gold 79	201 **Hg** mercury 80	204 **Tl** thallium 81	207 **Pb** lead 82	209 **Bi** bismuth 83	[209] **Po** polonium 84	[210] **At** astatine 85	[222] **Rn** radon 86
[223] **Fr** francium 87	[226] **Ra** radium 88	[227] **Ac*** actinium 89	[261] **Rf** rutherfordium 104	[262] **Db** dubnium 105	[266] **Sg** seaborgium 106	[264] **Bh** bohrium 107	[277] **Hs** hassium 108	[268] **Mt** meitnerium 109	[271] **Ds** darmstadtium 110	[272] **Rg** roentgenium 111							

1
H
hydrogen
1

Key

relative atomic mass
Atomic symbol
name
atomic (proton) number

Elements with atomic numbers 112–116 have been reported but not fully authenticated

* The lanthanoids (atomic numbers 58–71) and the actinoids (atomic numbers 90–103) have been omitted.
The relative atomic masses of copper and chlorine have not been rounded to the nearest whole number.

Physics formulae

There are a number of formulae that you will be expected to be able to use in your physics examination for Unit P2. However, you do not need to learn them by heart. You will be a given a formulae sheet in the examinations which will contain all the formulae from the unit. They are also shown below.

Specification statement	Equation
1.11	charge (coulomb, C) = current (ampere, A) × time (second, s) $Q = I \times t$
2.8	potential difference (volt, V) = current (ampere, A) × resistance (ohm, Ω) $V = I \times R$
2.15	electrical power (watt, W) = current (ampere, A) × potential difference (volt, V) $P = I \times V$
2.16	energy transferred (joule, J) = current (ampere, A) × potential difference (volt, V) × time (second, s) $E = I \times V \times t$
3.4	$\text{speed (m/s)} = \dfrac{\text{speed (m/s) = distance (m)}}{\text{time (s)}}$
3.5	$\text{acceleration (metre per second squared, m/s}^2) = \dfrac{\text{change in velocity (metre per second, m/s)}}{\text{time taken (second, s)}}$ $a = \dfrac{(v-u)}{t}$
3.13	force (newton, N) = mass (kilogram, kg) × acceleration (metre per second squared, m/s²) $F = m \times a$
3.14	weight (newton, N) = mass (kilogram, kg) × gravitational field strength (newton per kilogram, N/kg) $W = m \times g$
4.4	To calculate the momentum of a moving object: momentum (kilogram metre per second, kg m/s) = mass (kilogram, kg) × velocity (metre per second, m/s)
4.9	To calculate the change in momentum of a system: $\text{force (newton, N)} = \dfrac{\text{change in momentum (kilogram metre per second, kg m/s)}}{\text{time (second, s)}}$ $F = (mv - mu) / t$
4.10	work done (joule, J) = force (newton, N) × distance moved in the direction of the force (metre, m) $E = F \times d$
4.13	$\text{power (watt, W)} = \dfrac{\text{work done (joule, J)}}{\text{time taken (second, s)}}$ $P = \dfrac{E}{t}$
4.15	gravitational potential energy (joule, J) = mass (kilogram, kg) × gravitational field strength (newton per kilogram, N/kg) × vertical height (metre, m) $GPE = m \times g \times h$
4.16	kinetic energy (joule, J) = ½ × mass (kilogram, kg) × velocity² ((metre/second)² (m/s)²) $KE = \tfrac{1}{2} \times m \times v^2$

Index

Published by Pearson Education Limited, a company incorporated in England and Wales, having its registered office at Edinburgh Gate, Harlow, Essex, CM20 2JE. Registered company number: 872828

Edexcel is a registered trademark of Edexcel Limited

Text © Pearson Education Limited 2011

First published 2011

10 9 8 7 6 5

The rights of Mark Levesley, Aaron Bridges, Ann Fullick, Richard Grime, Miles Hudson, Penny Johnson, Sue Kearsey, Jim Newall, Damian Riddle, Sue Robilliard, Nigel Saunders, Mark Grinsell, Sue Jenkin, Ian Roberts, Julia Salter and David Swann to be identified as authors of this work have been asserted by them in accordance with the Copyright, Designs and Patents Act 1988.

British Library Cataloguing in Publication Data

A catalogue record for this book is available from the British Library

ISBN 978 184690 883 5

Typeset by eMC Design Ltd
Illustrated by Oxford Designers and Illustrators
Picture research by Louise Edgeworth, Susi Paz and Rebecca Sodergren
Printed in the UK by Scotprint

Acknowledgements

The publisher would like to thank all the teachers and students who helped in the development and trialling of this course, including:

Steven Rowe, Graham Hartland, David French and students at Tomlinscote School; Alex Dawes and students at the Jewish Free School; Suzanne Mycock, Sandra Fox and students at Chelmer Valley High School; David Liebeschuetz, Richard Brock, Elizabeth Andrews and the science team at Davenant Foundation School; Peter Bowen-Walker; Carol Chapman; Fay Dodds; Ben Lovick; Esther Ruston and Rupert Turpin.

We are grateful to the following for permission to reproduce copyright material:

Map on page 74 from Uko Gorter Natural History Illustration (www.ukogorter.com). Copyright of Uko Gorter, reproduced with permission; Graphs on page 78 from WHO Working Group on Infant Growth. An Evaluation of Infant Growth: a summary of analyses performed in preparation for the WHO Expert Committee on Physical Status: The use and interpretation of anthropometry. (WHO/NUT/94.8). Geneva. World Health Organization; 1994 p21. Copyright of World Health Organization, reproduced with permission; Graph on page 95 (Figure C) from Reduction of Serum Cholesterol with Sitostanol-Ester Margarine in a Mildly Hypercholesterolemic Population. (Miettinen, Puska, Gylling, Vanhanen, Vartiainen 1995.) 10.1056/NEJM199511163332002. Copyright of The New England Journal of Medicine, reproduced with permission; Graph on page 95 (Figure D) from Studies with Inulin-Type Fructans on Intestinal Infections, Permeability, and Inflammation (Francisco Guarner 2007) Journal of Nutrition vol. 137 no. 11. Copyright of Journal of Nutrition, reproduced with permission.

In some instances we have been unable to trace the owners of copyright material, and we would appreciate any information that would enable us to do so.

The author and publisher would like to thank the following individuals and organisations for permission to reproduce photographs:

(Key: b-bottom; c-centre; l-left; r-right; t-top)

2005 WFU/Ken Bennett: 2005 WFU / Ken Bennett 122; Alamy Images: Anthony Collins Cycling 240br, Archive Pics 32, Ashley Cooper pics 154tr, Bert Hoferichter 66, Bill Bachman 86, Darkened Studio 172, fStop 246br, imagebroker 240bl, INTERFOTO 30tr, ITAR-TASS Photo Agency 56, Lifestyle Concepts & Emotions 130, Marc Anderson 168, MartinShields 72tl, 72tr, Motoring Picture Library 248tr, Paul Dronsfield 52, Photofusion Picture Library 264br, RedFX 132tr, RIA Novosti 118, Rosemary Roberts 150, Sciencephotos 156bl, 209, Vario Images GmbH & Co.KG 222tr, World History Archive 30tl, Yoav Levy / PHOTOTAKE 186t; Art Directors and TRIP Photo Library: Andrew Lambert 214tr; BAA Aviation Photo Library: David J. Osborn 210bl; BBC Motion Gallery: 243; Richard Box: 206-207; Jacob A Bruinsma: 220; Caterham Cars: 235tl; Centre for Solar Energy Research, Glyndwr University at OpTIC: 214 (A); Phil Coomes: 265; Corbis: 224tr, Bettmann 158tr, Image Source 92, Ocean 222c, Tomas Rodriguez 220b; Courtesy Golden Rice Humanitarian Board. www.goldenrice.org: Courtesy Golden Rice Humanitarian Board. www.goldenrice. org 39; courtesy of Reaction Engines Ltd: 230, 230bl; © Crown Copyright / Office of Public Sector Information: http: / / www.direct.gov.uk / en / TravelAndTransport / Highwaycode / DG_070304 239; Dainese D-Air Racing system: 246tr; Deborah Metcalfe, Blue Eye FX Productions; Deborah Metcalfe, Blue Eye FX Productions; Deborah Metcalfe, Blue Eye FX Productions: Tom Smith for Blue Eye FX Productions 60t; DK Images: 76cl, Matthew Ward 218bc; Dr Mary Hebert at Newcastle Fertility Centre and Institute for Ageing & Health, Newcastle University: Dr Mary Hebert at Newcastle Fertility Centre and Institute for Ageing & Health, Newcastle University 44tl; Dr Panaviotis Zavos: Dr Panaviotis Zavos 44; Dreamstime.com: Artography 156tr; Lee Durant: 212; ACFSE - The Alliance for Consumer Fire Safety in Europe: 160tl, 160tr; Mary Evans Picture Library: 136tr, 272tr; Jonathon Fleetwood : 88, 90, 94t; Fotolia.com: Beelix 234br, jonnysek 248tl, Kramografie 248tc, youpee2305 234bl; Virgin Galactic: 224tr, Mark Greenberg 230tl; Getty Images: Bradley Kanaris 146tr, Brandon Cole / Visuals

Unlimited 74, Bruce Weaver / AFP 225, Chris Garrett / Stone 248cr, David McNew 44br, Dorling Kindersley 175, 194, Erik Von Weber / Stone 251, FPG / Hulton Archive 260, Getty 142t, ImageBank 180, JOE KLAMAR / AFP 60, Martin Bernetti 126, Matthew Stockman 61, Melissa McManus / Stone 58t, Mike Day 222br, Philip Evans / Flickr 157cr, Stuart McCall / Photographers Choice 247; Olivier Grunewald: 116-117; HotCan: HotCan 166tr; iStockphoto: aaM Photography, Ltd. 45, Mike DaBell 132bl, PicstoDisc 163cr; Joe Fox Photography: 238c; Lee Abley: 91cl, 91cr; Mark Levesley: Mark Levesley 50tr, 170c, Mark Levesley 50tr, 170c; LiaM: LiaM 138tr; Sam Loman: 26-27; Made With Molecules: Made With Molecules 140; Euro NCAP Communications Manager: 244tr; NASA: 228tr, HQ / GRIN 250cl, HQ-GRIN 228bl, 238tr, HQ-GRIN 228bl, 238tr, JPL / Caltech 250tr, JPL / University of Arizona 254, JPL-Caltech 231tr, Kennedy Space Center 256, MSFC 228br; Natural History Museum Picture Library: The Natural History Museum, London 76tl; Nature Picture Library: Asgeir Helgestad 266; Pearson Education Ltd: Trevor Clifford 267, Comstock Images 276cl; Pearson Education Ltd: PhotoDisk / David Buffington 82tr, Trevor Clifford 136cl, 139b, 154c, 155, 157cl, 178bl, 244bl, Creatas 166b, Photodisc. Chris Falkenstein 134, PhotoDisk 137, Debbie Rowe 120bl; Photo courtesy Lesley Pyke www. lesleypyke.com: Photo courtesy Lesley Pyke www.lesleypyke.com 142b; Photo courtesy of Texas A&M University College of Veterinary Medicine & Biomedical Sciences : Photo courtesy of Texas A&M University College of Veterinary Medicine & Biomedical Sciences 43tl, 43c, 43cl; Photolibrary.com: Fresh Food Images / Jason Lowe 34, Peter Arnold Images / Oldrich Karasek 271, Phototake Science / Yoav Levy 158bl, White / Karl Weatherly 58b; Photos.com: 151; Press Association Images: 42, 46, 48tr, 148, CHANDLER ALAN CHANDLER / PA Archive 226, CHARLES KRUPA / AP 268tr, DAMIAN DOVARGANES / AP 218tr, DEUTSCHE PRESS-AGENTUR / DPA 224cl, Matt Writtle / PA Archive 208; Reuters: China Daily China Daily Inform 38tr, Darrin Zammit Lupi 178tra, Jean-Paul Pelissier 263cl, Jo Yong hak 42tr, Radu Sigheti 37; Rex Features: 242tr, Action Press 264tr, Auto Express 235r, Giuliano Bevilacqua 234tr, KeystoneUSA-ZUMA 231br, 236tr, Mark Extance 70, MAURIZIO DI LORETI 270tr, Monkey Business Images 232, Neale Haynes 240tr, Offside 229, Shout 246bl, Sipa Press 258tr, Steve Meddle 164; Richard Martin: Richard Martin 54; Science Photo Library Ltd: A. Barrington Brown 36tr, 36l, 82cl, 262br, AJ Photo 85, Andrew Lambert Photography 146bl, 161, 219t, 219c, 219b, Antonia Reeve 83, Astier - Chru Lille 272br, Brian Bowes 42cl, CHARLES D. WINTERS 133, 139t, 156bc, 156br, 174tl, 186c, 186b, 196, 198, Cordelia Molloy 51, 62b, DAVID R. FRAZIER 274, DIRK WIERSMA 183, DR Gopal Murti 28bl, 50bl, DR M.A. Ansary 40, DR. John Brackenbury 220bl, E. R. DEGGINGER 138bl, 138br, Eye of Science 90bl, Hank Morgan 273, Health Protection Agency 270br, HR Bramaz / ISM 30cl, ISM 272cl, J. C. Revy / ISM 38bl, J.C. REVY, ISM 70cl, 70cr, John Durham 29, John Radcliffe Infirmary 49t, Josh Sher 263cr, Keith / Custom Medical Stock Photo 144, Kwangshin Kim 30cr, M.I. Walker 79, Martyn F. Chillmaid 166bl, 166br, 203, Maximilian Stock LTD 170t, 261, Mehau Kulyk 124, Peter Menzel 216, Michael Abbey 28tr, NASA 236br, 252, P.G. Adam, Publiphoto Diffusion 184t, Power and Syred 129, RIA Novosti 258cl, Robert Brook 184b, Simon Fraser 185, Ted Kinsman 248cl, Tony Camacho 131, Trevor Clifford Photography 170b, 218b, TRL LTD. 246bl, US Department of Energy 257; Sharon Loxton: 182; Shout Pictures: 242c; Shutterstock.com: Andreas Meyer 101, George Bailey 160c, Charlie Edward 162, Elena Moiseeva 149, Frannyanne 128, Gheorghe Bunescu Bogdan Mircea 174tr, Hywit Dimyadi 276tl, Jacob Hamblin 268br, János Németh 48bl, Li Wa 80, Marten Czamanske 163cl, Pichugin Dmitry 152, R. Gino Santa Maria 176, Rafa Irusta 280, Roberto Sanchez 48br, Sahua D 68, Sebastian Kaulitzki 49c, STILLFX 276tr, Vkrosh 276cr; Sylvania Lamps UK : Sylvania Lamps UK 156tl; Thinkstock: Hemera 62t, Photodisc 78; Professor Glenn R Gibson, Reading University: 94b; Wellcome Library, London: 262tr; Worldfirst racing car team: 64; www.unece.org/trans/danger/publi/ghs/ pictograms.html: Reproduced with the kind permission of the secretariat of the United Nations Economic Commission for Europe 154cr, 156cl, 156cr, 160bl, 160br, Y.Sugimoto / www.osaka-u.ac.jp: Y.Sugimoto / www.osaka-u.ac.jp 120tr

Cover images: Front: Shutterstock.com: Nadiya Sergey

All other images © Pearson Education

Every effort has been made to contact copyright holders of material reproduced in this book. Any omissions will be rectified in subsequent printings if notice is given to the publishers.